Study Guide
for Press and Siever's
Understanding Earth
SECOND EDITION

David M. Best

Northern Arizona University

W. H. Freeman and Company

New York

ACQUISITIONS EDITOR: Patrick Shriner
ASSISTANT EDITOR: Melina LaMorticella
PROJECT EDITOR: Erica Lane Seifert
PRODUCTION COORDINATOR: Paul Rohloff
EDITORIAL ASSISTANTS: Maia Holden, Jennifer King
COMPOSITION AND TEXT DESIGN: Christopher Wieczerzak
MANUFACTURING: Vail-Ballou

ISBN: 0-7167-2804-4

Printed in the United States of America

First Printing 1998

Contents

Preface

This *Study Guide* has been developed to help you excel in the introductory geology course using the textbook *Understanding Earth*, Second Edition, by Frank Press and Raymond Siever. It is designed to assist you in understanding the concepts in the textbook and also in getting the most out of the materials available to you on the CD-ROM that accompanies the textbook.

First there will be a description of the contents of each chapter of this Study Guide. Then there will be a brief discussion of some of the other materials available to you on the CD and the Web Site for *Understanding Earth*.

• Each chapter of the *Study Guide* begins with a **Chapter Summary** designed to help you better understand the coverage of the chapter. Each new topic is introduced after a diamond symbol. Within each chapter summary, boxed features (which appear in a two-column format) focus on topics that may require extra attention from you. Many of these focus on the role of "plate tectonics" in the chapter. As you will soon read, plate tectonics is a principle that is central to geology. Understanding its role is a good way to get a grip on understanding the whole course.

• The **Chapter Outlines** offer you a convenient breakdown of the coverage of each chapter. Using the outlines is an excellent way to organize your study and/or lecture notes.

• Most chapters then offer one or two exercises built around specific coverage within that chapter. These are meant to complement the **Interactive Geology Exercises** found on the CD. There are 34 of those on the CD covering topics in the 24 chapters of the book. You should work through both the printed and the electronic exercises in each chapter.

• There are about 15 **Practice Multiple-Choice Questions** for each chapter in the Study Guide. These should help you prepare for tests and also give you a good idea of how well you know a chapter. These represent but one of three practice multiple-choice quizzes available to you for each chapter. There is another—called **Q & A**—in each chapter on the CD. You will find yet another Q & A quiz for each chapter on the *Understanding Earth Web Site*. Each of the three practice quizzes are made up of questions that do not appear in the other two.

• **CD Resources** lists the materials available on each chapter of the CD. These include listing of the Interactive Geology Exercises, additional photos, and animations and videos.

• The **CD Lab** for each chapter is a short series of questions relating to materials on the CD. Your instructor may chose to assign some of these if there is a lab requirement to your course. The answers to these questions do not appear in this *Study Guide,* but the answers to the other exercises in this *Study Guide* do appear at the end of each chapter.

• The feature called **Final Review** is offered as way for you to make sure you know some basic points of the chapter. You should not feel comfortable about stopping your study of a chapter if you do not understand each point. However, you should also be aware that this is just one person's opinion of what you should you know. Your instructor may well have other points that he or she considers equally or more important.

OTHER HELP AVAILABLE TO YOU

As has been described above, there is at least one Interactive Geology Exercises for each chapter on the CD that accompanies the book, as well as an interactive Q & A multiple-choice self quiz that offers feedback on your answers and specific instructions for further study. There is an additional Q & A quiz for every chapter on the Understanding Earth Web Site at:

www.whfreeman.com/understandingearth

The CD Resource feature of this *Study Guide* lists a whole range of additional features on the CD that are designed to enlarge an understanding of the topics presented in the textbook. The best way for you to discover what these are is to explore the features using the actual CD.

The *Web Site* is designed as an ever-changing, ever-growing resource for students and instructors. Like the CD, it is an excellent way to explore the wonders of geology.

A NOTE ON THIS REVISION

The author of this *Study Guide,* David M. Best, has been greatly aided in this revision by the team who worked on the CD—especially the CD's development editor, Diana Siemens. Likewise, the CD—especially its Interactive Geology Exercises—reflects the learning ideas of Dr. Best as demonstrated in many of the textbook figure exercises that first appeared in the First Edition *Study Guide,* but now appear in an electronic form on the CD. We think that you will find that the combination of the textbook, the CD, and the *Study Guide* (while not neglecting the crucial role of your instructor's lectures) is the best possible way to learn the topic of physical geology, do well in this course, and have some genuine enjoyment while doing so.

(The publisher would be very interested in hearing what you think of this package of learning tools. If you would care to comment we pledge that we will keep your ideas in mind as we prepare the next edition package. Send any comments by email to {cdrom@whfreeman.com}. Thank you.)

Good luck!

1

Building a Planet

Geology, the science that studies Earth, is sometimes considered a hybrid science because it draws from the basic ideas of chemistry, physics, mathematics, and biology. Like other scientists, geologists use the scientific method. They observe the natural world and record their observations. They develop hypotheses to explain the things they have observed, then test these hypotheses to see if they can be confirmed by further observations. When a hypothesis is confirmed repeatedly by other scientists' experiments, it may become a theory. Although a theory is never absolutely provable, confidence grows in theories that withstand repeated tests and are able to predict the results of new experiments. For example, geologists have confidence in the principle of uniformitarianism, which states that the processes that have shaped Earth have not changed over geologic time.

Our solar system, with its 9 planets and the more than 60 moons that orbit them, developed about 4.5 billion years ago. Scientists believe that the solar system formed from a rotating cloud of gas and fine dust, called a nebula. The nebula contracted under the force of gravity, and this contraction accelerated the rotation of the particles and flattened the cloud into a disk. Gravity pulled the particles toward the center, where they accumulated into a proto-Sun. As the material in the proto-Sun was compressed under its own weight, it became hot and dense enough to begin nuclear fusion, which continues to this day.

◆   Although most of the matter in the original nebula was concentrated in the proto-Sun, a disk of gas and dust called the solar nebula remained around it. The solar nebula heated up as it flattened into a disk, then began to cool. Many of the gases condensed and clumped together as the nebula cooled. These clumps, called planetesimals, created larger bodies as they continued to collide and stick together. Finally, a few of these larger bodies swept up the others to form our nine planets in their present orbits.

The inner planets—Mercury, Venus, Earth, and Mars—are small and rocky. Radiation and matter streaming from the Sun blew away most of their hydrogen, helium, water, and other light gases and liquids, leaving behind dense metals such as iron and other heavy, rock-forming substances. The giant outer planets—Jupiter, Saturn, Uranus, and Neptune—were big enough and their gravitational attraction was strong enough to hold on to the lighter (and by now cooler) nebular constituents. That is why they are composed mostly of hydrogen and helium and the other light constituents of the original nebula.

♦   Soon after Earth aggregated into a rocky mass, it is likely that it collided with a planetesimal the size of Mars. The heat energy from this giant impact melted the Earth and began the process of differentiation. The Moon formed from the debris left over from this impact.

♦   Differentiation led to the formation of Earth's crust and, later, the continents. Iron sank to the center, and lighter material such as gases floated upward to form a crust. A mantle composed of rock of intermediate density formed between the crust and the mantle.

The growth of continents began soon after differentiation, and it has continued since. Scientists think that magma floated up from the molten interior of Earth to the surface, where it cooled and solidified to form a rocky crust. With repeated melting and solidification, the lighter materials separated from the heavier ones and floated to the top to form the primitive nucleus of the continents. Rainwater and other components of the atmosphere decomposed and disintegrated the rocks to decompose and disintegrate. Water, wind, and ice then loosened and moved rocky debris to low-lying places, where it accumulated in thick layers, forming beaches, deltas, and the floors of adjacent seas. As this process was repeated through countless cycles, continents formed.

♦   Most geologists believe that Earth's oceans and atmosphere formed from water and gases that boiled off during heating and differentiation. As Earth heated and its materials partially melted, water vapor and other gases were freed and carried to the surface by magmas and released through volcanoes. The earliest atmosphere was entirely different from our current atmosphere, which is composed primarily of nitrogen and oxygen. The production of significant amounts of free oxygen and its persistence in the atmosphere probably resulted from the photosynthesis of early one-celled plants.

♦   All the terrestrial planets have differentiated, although their evolutionary paths have differed. *(See the Planetary Geology expansion module on the CD-ROM for more information about the planets and moons in our solar system.)*

♦   Plate tectonics, a concept that revolutionized geology in the 1960s, is the idea that Earth's behavior is caused largely by the formation, movement, interactions, and destruction of the large rigid plates found on Earth's surface. *(See the accompanying article for a summary of this important theory.)*

THE THEORY OF PLATE TECTONICS

There is no more important concept in the field of geology than the theory of plate tectonics. As this course develops, we will see that there are many subspecialties in geology, such as sedimentary geology (the study of sediments and sedimentary rocks), metamorphic petrology (the study of metamorphic rocks, those produced by changes in heat, pressure, and chemistry), and geophysics (the study of Earth's physical properties by quantitative methods). These three major areas of study help explain the concept of plate tectonics, and, in turn, plate tectonics helps explain a great deal about the formation of sediments and metamorphic rocks and about the physical changes that take place on Earth's surface. Specialists in virtually all the areas of geology have incorporated the theory of plate tectonics into their studies of Earth.

Chapter 1 introduces the basic ideas of plate tectonics in terms of the boundaries where plates converge, separate, and slip past one another as they move about on Earth's surface. At divergent boundaries, new lithosphere forms from magma rising through a rift at a mid-ocean ridge. At convergent boundaries, where plates collide, one plate sinks beneath the other in a process called subduction. In subduction zones, lithosphere is consumed by being returned to the mantle. Plates slide past each other at transform fault boundaries, and their motion often results in earthquakes.

As you progress through the textbook, it will become evident that plate tectonics plays a major role in a number of geological processes: volcanism, earthquakes, deformation of the continents, landscape evolution, and sedimentation, to name a few. Each of these topics is covered in considerable detail in later chapters. One of the beauties of the plate-tectonics theory is that it provides a good explanation for so many different processes observed in the geological world.

♦ Geologists choose to concentrate on different aspects of geology. Some are interested in geology as a pure science; others want to use their knowledge in practical ways. Scholarly geologists may be teachers and researchers in universities or work for government agencies or private firms that engage in construction or draw materials from the Earth. Some geologists are primarily interested in advancing our understanding of processes that can result in natural disasters. Others devote their studies to protecting the environment. Economic geologists search out the remaining undiscovered deposits of coal, oil, natural gas, and the ores from which our metals and industrial chemicals are derived.

CHAPTER 1 OUTLINE

PLANETARY DIFFERENTIATION

Fill in the blanks in the following description of planetary differentiation. The numbers in the description correspond to the numbers shown in the figures below the description. **Study Hint:** *If you don't do well on this exercise, you should review the textbook's discussion of this topic on pages 9 and 10.*

Early Earth was probably a homogenous mixture with no continents or oceans. In the process of differentiation, (1) _____ sank to the center and lighter material (represented by the squiggled lines in b below). As a result, Earth is a zoned planet (c). At the center is a (2) _____. Surrounding that is a (3) _____. The next zone is called the (4) _____. The outermost zone is called the (5) _____.

(a)   (b)   (c)

PRACTICE MULTIPLE-CHOICE QUESTIONS

Circle the option that best answers the question.

1. A hypothesis that has accumulated a considerable body of support is called:
    (a) A principle
    (b) A proved hypothesis
    (c) A theory
    (d) The scientific method

2. Which of the following statements best describes current scientific thinking about the beginning of the universe?
   (a) Matter existed as dust and clouds that began to solidify on their own about 4.5 billion years ago.
   (b) A superdense point of matter and energy exploded between 15 and 20 billion years ago, expanding into what we know as the universe.
   (c) Several points of superdense matter and energy that existed throughout space exploded simultaneously, filling the universe with dust and clouds.
   (d) Matter in our solar system was drawn into outer space as the result of gravitational attraction.

3. As the proto-Sun began to form, which of the following changes took place?
   (a) Only certain elements were drawn into its interior by selective gravitational attraction.
   (b) Nuclear fusion, resulting from extremely high temperatures and pressures, transformed hydrogen atoms into helium atoms.
   (c) Material was thrown out immediately because of the liquid nature of the hydrogen present.
   (d) All the mass imploded into the proto-Sun.

4. Which of the following statements best describes the composition of the inner and outer planets of our solar system?
   (a) The inner planets are small and rocky because they formed close to the Sun, where conditions were so hot that gases could not be retained.
   (b) The inner planets and outer planets have the same basic composition.
   (c) The giant outer planets have cores composed of gases.
   (d) The outer planets are larger, which implies that their composition is different.

5. The Earth began to differentiate into a core, mantle, and crust:
   (a) About 4.6 billion years ago
   (b) About 4.6 million years ago
   (c) Between 4.1 and 4.4 billion years ago
   (d) Between 4.1 and 4.4 millon years ago

6. Earth's oceans and atmosphere first formed from gases:
   (a) Brought to Earth from the outer planets by comets
   (b) Released by volcanoes some 4 billion years ago
   (c) Produced by photosynthesis of plants
   (d) Produced by melting of the lithosphere

7. The lithosphere can be characterized as:
   (a) Composed of partially molten rock and supporting the asthenosphere
   (b) Having a thickness of more than 200 km
   (c) Being solid and riding atop the asthenosphere
   (d) A continuous shell that surrounds the outermost shell of the Earth

8. Earth's lithospheric plates move because:
   (a) Convection allows hot matter to rise and cool matter to sink
   (b) They float atop the world's oceans
   (c) They float on Earth's liquid mantle
   (d) The asthenosphere is cooler than the mantle

9. An excellent example of a convergent plate boundary can be found:
   (a) Along the western coast of South America
   (b) In the middle of the Atlantic Ocean
   (c) Along the San Andreas fault in southern California
   (d) In central Kansas

10. The Mid-Atlantic Ridge is an example of a:
   (a) Convergent boundary
   (b) Divergent boundary
   (c) Transform boundary
   (d) Mid-plate hot spot

11. New oceanic crust is created at:
   (a) Convergent boundaries
   (b) Subduction zones
   (c) Transform boundaries
   (d) Divergent boundaries

12. The book's description of the origin and early evolution of Earth is:
   (a) A proved fact
   (b) A theory
   (c) A hypothesis
   (d) Pure guesswork

13. The study of natural disasters that occur on Earth includes examining:
   (a) Landslides, volcanic eruptions, earthquakes, and floods
   (b) The pollution of groundwater by toxic chemical spills
   (c) The production of acid rain from fossil fuel combustion
   (d) Calculations of the consumption of oil and coal reserves

CD RESOURCES

SUPPLEMENTAL DIAGRAMS

1.21   Spatial and temporal scales
1.22   Three mechanisms that would cause early Earth to heat up
1.23   Olympus Mons compared to other mountains
1.24   Divergent boundary and seafloor spreading
1.25   A rift zone and a subduction zone
1.26   Convection and convergent and divergent plate boundaries
1.27   Pangaea
1.28   Computer-generated best-fit union of continents
1.29   Earth's plates today

SUPPLEMENTAL PHOTOS

1.30   View from Pasadena, California, on a clear day
1.31   View from Pasadena, California, on a smoggy day

FIRST EDITION SLIDE SET

1.1   The Earth
1.2   Space shuttle view of Earth's atmosphere
1.3   Space shuttle view of Himalayas
1.4   Relief of Earth's surface

SPACEVIEWS

SV1.1   Earth rising over Moon as seen from *Apollo 11*
SV1.2   *Apollo 12* Earthrise over Moon
SV1.3   Impact site on Jupiter of Comet Shoemaker-Levy 9
SV1.4   Fragment Q impacts on Jupiter from Comet Shoemaker-Levy 9
SV1.5   Water ice clouds over western flank of Ascreaus Mons
SV1.6   Jupiter, *Voyager 1*
SV1.7   Water ice (frost) on Martian soil and rocks
SV1.8   Saturn, *Voyager 1*
SV1.9   Surface of Venus, *Magellan*

EXPANSION MODULE: PLANETARY GEOLOGY

36.1   Stony meteorite
36.2   The Barringer Crater
36.3   Detail of a lunar crater
36.4   Mare Imbrium and surrounding highlands
36.5   Mare basalt
36.6   Impact breccias

ANIMATIONS AND VIDEOS

ILLUSTRATED GLOSSARY

EXERCISES

Three Types of Plate Boundaries

TOOLS

Geologic Time Scale

CD LAB

1. Which planet has huge shield volcanoes, the largest in the solar system?
    (Hint: See the Planetary Geology expansion module.)

2. The Moon is about:
    (a) 1/2 the size of Earth
    (b) 1/4 the size of Earth
    (c) 1/8 the size of Earth
    (Hint: See the Planetary Geology expansion module.)

3. Triton is a moon of:
    (a) Jupiter
    (b) Saturn
    (c) Uranus
    (d) Neptune
    (Hint: See the Planetary Geology expansion module.)

4. Is the Holocene epoch earlier or later than the Miocene epoch?
    (Hint: See the Geologic Time Scale.)

5. Which of the following is *not* a hypothesized mechanism by which early Earth heated up?
    (a) Impact of a giant bolide
    (b) Plate tectonics
    (c) Disintegration of radioactive elements
    (d) Compression of the planet under its own weight
    (Hint: See the supplemental diagrams for Chapter 1.)

6. What is the fundamental factor that makes life possible on Earth?
    (a) The distance of Earth from the Sun
    (b) The oceans
    (c) The atmosphere
    (d) Liquid water
    (Hint: See the first edition Slide Set Lecture Notes.)

7. Plate tectonics on Earth began as early as:
    (a) 4 billion years ago
    (b) 3 billion years ago
    (c) 2 billion years ago
    (d) 100 million years ago
    (Hint: See the Geologic Time Scale, Proterozoic era.)

FINAL REVIEW

After reading this chapter, you should:

- Understand the scientific method and the difference between hypothesis, theory, and fact
- Understand the principle of uniformitarianism
- Know the nebular hypothesis of the Sun's origin
- Know the text's explanation of the origin of the planets in our solar system
- Understand what is meant by planetary differentiation
- Know the text's explanation of the formation of continents, oceans, and the atmosphere
- Thoroughly understand the chapter's explanation of plate tectonics
- Know the differences between the three types of plate boundaries
- Broaden (or begin to broaden) your notion of time in reference to geologic time

ANSWER KEY

PLANETARY DIFFERENTIATION

1. Iron, 2. Solid iron inner core, 3. Liquid iron outer core, 4. Mantle, 5. Crust

PRACTICE MULTIPLE-CHOICE QUESTIONS

1. c, 2. b, 3. b, 4. a, 5. c, 6. b, 7. c, 8. a, 9. a, 10. d, 11. d, 12. c, 13. a

Minerals: Building Blocks of Rocks

Geologists define a mineral as a naturally occurring, inorganic, solid crystalline substance with a specific chemical composition. Minerals are the building blocks of rocks; some rocks consist of only one mineral, others of several different minerals. Minerals, in turn, are composed of one or more chemical elements.

◆   In 1805, John Dalton, an English chemist now considered the father of modern atomic theory, hypothesized that the various chemical elements consist of different kinds of atoms, that all atoms of any given element are identical, and that chemical compounds are formed by combining atoms of different elements in definite proportions. By the early twentieth century, the scientific community, building on Dalton's ideas, had reached a basic understanding of the structure of matter: An atom is the smallest unit of an element that retains the physical and chemical properties of that element. *(See the accompanying article for a summary of the text material on the atomic structure of matter and a discussion of chemical reactions.)*

◆   The periodic table (Figure 2.6) shows the elements (from left to right) in order of atomic number (number of protons). All the elements in the leftmost column have a single electron in their outer shells and have a strong tendency to lose that electron in chemical reactions. Elements in the second column from the left have two electrons in their outer shells and a strong tendency to lose them both in chemical reactions. Toward the right side of the table, the two columns headed by oxygen (O) and fluorine (F) group the elements that tend to gain electrons in their outer shells. The columns between the two on the left and the two headed by oxygen and fluorine have varying tendencies to gain, lose, or share electrons. The elements in the last column on the right, headed by helium (He), have full outer shells and thus no tendency either to gain or to lose electrons.

ATOMIC STRUCTURE

The concepts presented here are basic to understanding atoms and elements and how atoms form chemical compounds. An atom consists of a nucleus containing protons and neutrons surrounded by an electron shell. The nucleus makes up the atom's mass (also sometimes called the atomic weight). The atomic number of an element is the number of protons in the nucleus, and this number does not change. In an atom that is electrically neutral, the number of protons equals the number of electrons; the positive charges of the protons offset the same number of negative charges of the electrons. It is possible to change the number of neutrons in the nucleus of an atom; the result is called an isotope. Although additional neutrons do not change the atom's charge balance, they do alter the element's atomic mass.

Two or more elements can combine in a chemical reaction to form a chemical compound. During a chemical reaction, electrons in the outer shells of the constituent elements interact. Because the number of protons in an element does not change, when an electron in the outer shell is either lost or gained, the atom becomes electrically charged and is called an ion. If it loses an electron, it becomes positively charged (and is called a cation); if it gains an electron, it becomes negatively charged (and is called an anion). Because cations and anions possess an electrical charge, they have a tendency to link up with each other (opposite charges attract). This linking fills the outer electron shells of the atoms, resulting in an electrically neutral compound. This type of linking is called ionic bonding; the text uses sodium chloride (NaCl) as a classic example.

Elements can also combine chemically by sharing electrons instead of gaining or losing them. This type of bonding is called covalent bonding. An example is methane. Four hydrogen atoms react with one carbon atom; the carbon atom has four electrons in its outer shell, and each hydrogen atom has one. When these electrons are shared, all five atoms act as if each had a full complement of eight electrons in its outer shell. *(See the Chapter 2 supplemental diagrams on the CD for an illustration of the covalent bonding of methane.)*

Some complex minerals have a covalent bond holding together the anion complex and an ionic bond linking the anion complex to a cation. The mineral calcite, $CaCO_3$, has a covalent bond holding together the carbonate complex, $(CO_3)^{2-}$ (an anion), and an ionic bond linking the carbonate complex to the Ca^{2+} cation.

♦   The atoms in a mineral are arranged in an ordered three-dimensional array, or crystal. Crystallization starts with the formation of microscopic single crystals whose boundaries are natural flat surfaces, called crystal faces. A mineral's crystal faces express its internal atomic structure. As a mineral crystallizes, the initial crystals grow larger, maintaining their crystal faces as long as they are free to grow. Large crystals with well-defined faces form when growth is slow and steady and there is enough space to allow growth without interference from other crystals nearby. Most large mineral crystals form in open spaces in rocks.

Crystallization can begin in several ways. Freezing a liquid is one way. For example, magma—a hot, molten liquid rock—crystallizes solid minerals when it cools. Another way is by precipitation as liquids evaporate from a solution. For example, deposits of halite (table salt) can form when seawater evaporates in a shallow bay. Crystals also form when atoms and ions in solids become mobile and rearrange themselves at high temperatures. Mica forms in this way.

Two major factors control the arrangement of atoms and ions in a crystal structure: the number of neighboring atoms or ions, and their size. Ion size is related to the atomic structures of the elements, and it increases with the number of electrons and electron shells. An ion's charge also affects its size; most cations are small, and most anions are large. For this reason, most of the space of a crystal is occupied by the anions, and cations fit into the spaces between them. Cations of similar sizes and charges tend to substitute for one another and to form compounds with the same crystal structure but differing chemical compositions. Cation substitution is common in silicate minerals.

The same combinations of elements in the same proportions can sometimes form more than one kind of crystal structure and therefore more than one kind of mineral. These alternative possible structures for a single chemical compound are called polymorphs. For example, diamond and graphite are polymorphs of carbon.

◆   Minerals are classified according to their chemical composition. Table 2.1 shows six mineral classes. There are about 30 rock-forming minerals; the most common are silicates, carbonates, oxides, sulfides, and sulfates.

Silicates are the most abundant minerals in Earth's crust. The basic building block of all silicate minerals is the silicate ion, which takes the form of a four-sided pyramid called a tetrahedron. Tetrahedra may be isolated or linked in rings, single chains, double chains, sheets, or frameworks.

Carbonates contain the carbonate ion, which consists of a carbon atom surrounded by three oxygen atoms in a triangle. Calcite is one of the most common carbonate minerals. Groups of carbonate ions are arranged in sheets somewhat like the sheet silicates and are bonded by layers of cations.

Oxides are compounds in which oxygen is bonded to atoms or cations of other elements, usually metallic ions such as iron. Most oxide minerals are ionically bonded, and their structures vary with the size of the metallic cations.

The chief ores of many valuable minerals, such as copper, zinc, and nickel, are sulfides. The sulfide ion consists of a sulfur atom that has gained two electrons in its outer shell. These minerals have diverse structures, depending on how the sulfide anions combine with the metallic cations.

In sulfates, sulfur is present as the sulfate ion, a tetrahedron composed of one sulfur atom that has lost six electrons from its outer shell, joined with four oxygen ions. The sulfate ion is the basis for a variety of structures.

♦   To understand the geological origin of rocks, geologists must know the compositions and structures of their constituent minerals. In the field, geologists rely on chemical and physical properties that can be observed relatively easily. The major physical properties include hardness, cleavage, fracture, luster, color, specific gravity and density, and crystal habit.

Hardness is a measure of the ease with which the surface of a mineral can be scratched. Within groups of minerals having similar crystal structures, increasing hardness is related to factors that also increase bond strength.

Cleavage is the tendency of a crystal to break along flat planar surfaces, and it decreases with bond strength—if bond strength is high, cleavage is poor; if bond strength is low, cleavage is good. Cleavage is classified according to two primary sets of characteristics: the number of planes and pattern of cleavage, and the quality of surfaces and ease of cleaving.

Fracture is the tendency of a crystal to break along irregular surfaces other than cleavage planes. Conchoidal fractures show smooth, curved surfaces like those of a thick piece of broken glass. Other fractures are fibrous or splintery.

Luster refers to the way the surface of a mineral reflects light. Ionically bonded crystals tend to be glassy; covalently bonded minerals are more variable. Many minerals with covalent bonds have an adamantine luster, like that of diamond. Metallic luster is shown by pure metals, such as gold, and sulfides, such as galena. Pearly luster is the result of reflections of light from planes beneath the surfaces of translucent minerals. Mother-of-pearl (aragonite) has a pearly luster.

The color of a mineral is imparted by light that is either transmitted through or reflected by crystals, irregular masses, or a streak. Streak is the color of the fine deposit of mineral dust left on an abrasive surface when a mineral is scraped across it. Color is determined both by the kinds of ions found in the pure mineral and by trace impurities.

Density is the mass per unit volume of a mineral. A standard measure of density is specific gravity, which is the weight of a mineral in air divided by the weight of an equal volume of pure water at 4°C. Density depends on the atomic weight of the mineral's ions and on how closely they are packed in the mineral's crystal structure.

A mineral's crystal habit is the shape in which its individual crystals or aggregates of crystals grow. Crystal habits are often named after common geometric shapes, such as blades, plates, and needles.

CHAPTER 2 OUTLINE

MAJOR MINERAL GROUPS

In the blank to the left of the mineral name, classify each mineral using one of the major mineral groups listed below.

_____ 1. Amphibole	_____ 8. Feldspar	_____ 15. Mica
_____ 2. Anhydrite	_____ 9. Galena	_____ 16. Olivine
_____ 3. Beryl	_____ 10. Garnet	_____ 17. Pyrite
_____ 4. Calcite	_____ 11. Graphite	_____ 18. Pyroxene
_____ 5. Clay mineral	_____ 12. Gypsum	_____ 19. Quartz
_____ 6. Diamond	_____ 13. Halite	_____ 20. Spinel
_____ 7. Dolomite	_____ 14. Hematite	_____ 21. Talc

A. Silicate	D. Sulfate	G. Halide
B. Oxide	E. Native element	
C. Sulfide	F. Carbonate	

PRACTICE MULTIPLE-CHOICE QUESTIONS

Circle the option that best answers the question.

1. Which of the following is an accurate statement about the structure of atoms?
    (a) Protons move around the nucleus of an atom in a series of orbits, each of which can hold up to eight electrons.
    (b) The number of electrons in an atom is equal to the number of protons plus the number of neutrons.
    (c) Atoms of most elements can gain or lose electrons from their outer shells to form ions.
    (d) The positively charged cation of an element is heavier than the atom of that element, because of the addition of electrons.

2. Ionic bonds are made by:
    (a) Electrical attraction between ions produced by the transfer of electrons between atoms, from cations to anions
    (b) The sharing of some electrons in such a way that they orbit around both nuclei
    (c) Electrical attraction between ions that have opposite electrical charges
    (d) Electrical attraction between ions with the same electrical charge, but very different sizes

3. An isotope of an element is an atom that contains:
    (a) Additional electrons, which are placed in shells closest to the nucleus
    (b) One more proton than the original atom of the element
    (c) Fewer electrons than the original atom of the element
    (d) One or more additional neutrons, which increase the atomic mass of the element

4. The gain or loss of electrons transforms an atom into:
    (a) An ion
    (b) A nucleus
    (c) A complex ion
    (d) An isotope

5. The sharing of electrons by adjacent atoms is known as:
    (a) Ionic bonding
    (b) Covalent bonding
    (c) Tetrahedral bonding
    (d) Crystallization

6. Cation substitution is common in:
    (a) Sulfides
    (b) Oxides
    (c) Halides
    (d) Silicates

7. Polymorphs are minerals that:
    (a) Have very different chemical compositions
    (b) Have the same chemical compositions but different crystal structures
    (c) Differ because of cation substitution
    (d) Exhibit the same physical properties

8. Minerals are classified into groups based on their:
    (a) Cations
    (b) Anions
    (c) Hardness
    (d) Color

9. All silicate minerals contain the same basic building block, which is:
    (a) At least one chain of silica tetrahedra
    (b) A double chain of silica tetrahedra
    (c) A framework structure, such as that seen in quartz
    (d) The silica tetrahedron

10. The type of bond that holds together the silica tetrahedra found in single chain, double chain, and sheet silicates is:
    (a) Ionic
    (b) Covalent
    (c) Tetrahedral
    (d) A combination of ionic and covalent

11. A mineral's hardness is generally attributable to its:
    (a) Temperature of formation
    (b) Cleavage
    (c) Bond type
    (d) Atomic weight

12. An adamantine luster would be shown by:
    (a) Diamond
    (b) Aragonite (mother of pearl)
    (c) Gold
    (d) Galena

CD RESOURCES

SUPPLEMENTAL DIAGRAMS

2.26   A sodium atom and a sodium ion
2.27   Electron sharing in methane
2.28   Carbonate ion
2.29   Phosphate ion
2.30   Sulfate ion
2.31   How atoms combine to form rocks

SUPPLEMENTAL TABLE

2.6   Relative abundance of the ten most common elements in Earth's crust

SUPPLEMENTAL PHOTOS

2.32   Fluorite crystal structure.
2.33   Pyroxenes: augite (left) and diopside (right)
2.34   Amphibole: two samples of hornblende.
2.35   Mineral properties: hardness
2.36   Mineral properties: fracture (obsidian)
2.37   Mineral properties: luster (hematite and pyrite)
2.38   Mineral properties: luster (sphalerite, calcite, kaolinite, anglesite)
2.39   Mineral properties: color.

FIRST EDITION SLIDE SET

Slide 2.1   Aragonite forming in a cave
Slide 2.2   Four-atom clusters of sulfur atoms
Slide 2.3   X-ray diffraction powder pattern of quartz
Slide 2.4   Calcite cleavage
Slide 2.5   Diamond and graphite; carbon polymorphs
Slide 2.6   Zoned tourmaline
Slide 2.7   Polished malachite

SECOND EDITION SLIDE SET

Slide 2.1   Comparison of the X-ray diffraction powder patterns of halite, galena, and quartz.

ILLUSTRATED GLOSSARY

EXERCISES

Elements of Common Minerals
The Periodic Table
Ionic Bonding

TOOLS

Mineral Databank

CD LAB

1. What is the 10th most abundant element in Earth's crust?

    (Hint: See Table 2.6 in the Chapter 2 tables.)

2. What is the chemical formula for methane?

    (Hint: See the Chapter 2 supplemental diagrams.)

3. Sapphires are actually the mineral:
    (a) Beryl
    (b) Corundum
    (c) Garnet
    (d) Anglesite

    (Hint: See the Chapter 2 supplemental photos.)

4. Which of the following minerals has a dull luster?
    (a) Kaolinite
    (b) Sphalerite
    (c) Calcite
    (d) Anglesite

(Hint: See the Chapter 2 supplemental photos.)

5. Crystals of calcium carbonate are likely to form:
    (a) On the deep ocean floor
    (b) Near a convergent plate boundary
    (c) In a limestone cave
    (d) In glaciated regions

(Hint: See the first edition Slide Set.)

6. What is the hardness of corundum?
    (a) 4.5
    (b) 5
    (c) 8.1
    (d) 9

(Hint: See the Mineral Databank tool.)

7. What kind of fracture does silver display?
    (a) Conchoidal
    (b) Hackly
    (c) Uneven
    (d) Splintery

(Hint: See the Mineral Databank tool.)

8. You need to identify an unknown mineral sample. The sample is pale yellow with a metallic luster and has a hardness of 6.5. Which of the following minerals is this sample likely to be?
    (a) Sulfur
    (b) Beryl
    (c) Pyrite
    (d) Garnet

(Hint: See the Mineral Databank tool.)

FINAL REVIEW

After reading this chapter, you should:

- Be able the recite the geological definition of a mineral
- Understand the configuration of atoms and know what isotopes are
- Understand why electrons are gained and lost during chemical reactions and the results of these gains and losses
- Be able to go to the periodic table of the elements and give an explanation of its structure (Figure 2.6, p. 33)
- Understand the formation of chemical bonds
- Understand the basics of crystallization
- Know what cation substitution is
- Be able to describe the five major rock-forming mineral groups (silicates, carbonates, oxides, sulfides, and sulfates) discussed in the text
- Be able to list the seven physical properties of minerals and understand what each means (Table 2.5, p. 51)

ANSWER KEY

MAJOR MINERAL GROUPS

1. A, 2. D, 3. A, 4. F, 5. A, 6. E, 7. F, 8. A, 9. C, 10. A, 11. E, 12. D, 13. G, 14. B, 15. A, 16. A, 17. C, 18. A, 19. A, 20. A, 21. A

PRACTICE MULTIPLE-CHOICE QUESTIONS

1. c, 2. a, 3. d, 4. a, 5. b, 6. d, 7. b, 8. a, 9. d, 10. d, 11. c, 12. a

Rocks: Records of Geologic Processes

Understanding rock properties and using these properties to deduce a rock's geologic origins is the primary aim of a geologist, and this chapter gives an overview of how geologists interpret the three great families of rock—igneous, sedimentary, and metamorphic. The appearance, texture, mineralogy, and chemical composition of a rock reveal how and where it formed. Rock patterns found in subsurface drilling and in outcrops can help us reconstruct geologic history. The rock cycle is the set of processes that convert each type of rock into the other two, and all these processes are driven by plate tectonics.

◆ Igneous rocks crystallize from a magma, a mass of melted rock that originates deep in the crust or upper mantle. There are two major types of igneous rocks: intrusive and extrusive. Intrusive igneous rocks are formed by slowly crystallizing magmas that have intruded rock masses deep in Earth's interior. They have interlocking large crystals, which grew slowly as the magma gradually cooled. Extrusive igneous rocks form from rapidly cooled magmas that erupt at the surface. They are easily recognized by their glassy or fine-grained texture.

Most of the minerals of igneous rocks are silicates. The common silicate minerals found in igneous rocks include quartz, feldspar, mica, pyroxene, amphibole, and olivine.

◆ Sedimentary rocks are composed of sediments—loose particles such as sand, silt, and the shells of organisms. The solid fragments of rock produced by weathering are called clastic particles. Clastic particles are laid down as sediment by running water, wind, and ice. The dissolved products of weathering are ions or molecules in the waters of soils, rivers, lakes, and oceans. These dissolved substances are precipitated from water by chemical and biochemical reactions and accumulate as layers of chemical and biochemical sediments.

◆ The process of lithification converts sediments into solid rock in one of two ways: by compaction, as grains are squeezed together by the weight of overlying sediment into a mass denser than the original, and by cementation, as minerals precipitate around deposited particles and bind them together.

Sedimentary rocks cover much of Earth's land surface and seafloor. Although most rocks found on the surface are sedimentary, they form only a thin layer atop the igneous and metamorphic rocks that make up the main volume of the crust.

Clastic sediments are composed mostly of silicate minerals, typically quartz, feldspar, and clay minerals. The most abundant minerals of chemically or biochemically precipitated sediments are carbonates.

◆ Metamorphic rocks are produced when heat and pressure deep in the Earth cause any kind of rock—igneous, sedimentary, or other metamorphic rock—to change its mineralogy, texture, or chemical composition while maintaining its solid form. Regional metamorphism occurs where high pressures and temperatures extend over large regions, such as at convergent plate boundaries. Contact metamorphism occurs where high temperatures are restricted to smaller areas, such as the rocks near and in contact with an intrusion.

Metamorphic rocks are composed mainly of silicates, because these rocks are transformations of other rocks that are rich in silicates. Typical minerals include feldspar, mica, pyroxene, and amphibole, the same kinds of silicates found in igneous rocks. Several other silicates—kyanite, staurolite, and some varieties of garnet—form under conditions of high pressure and temperature in a crustal setting and are not characteristic of igneous rocks. These minerals are good indicators of metamorphism.

◆ The rock cycle, shown in Figure 3.10 on page 69, is a set of geologic processes by which each of the three groups of rocks is formed from the other two. The text presents an example to illustrate the various parts of the rock cycle. We can start with a plutonic episode, where preexisting rocks deep in Earth's interior melt to form igneous rocks. At plate collision boundaries, igneous rocks are uplifted as a high mountain chain. Weathering and erosion then create sediments, which are deposited, buried, and lithified. As the sedimentary rocks are buried more deeply, metamorphism takes place. With further heat and pressure, the rocks melt and form a new magma from which igneous rocks will crystallize, beginning the cycle again.

◆ Plutonism, volcanism, tectonic uplift, metamorphism, weathering, sedimentation, transportation, deposition, and burial are the geologic processes that combine in the rock cycle to convert the three groups of rocks to one another.

These processes are all driven by plate tectonics. *(See the accompanying article and the Rock Cycle tool on the CD for more information about how plate tectonics relates to the rock cycle.)*

PLATE TECTONICS AND THE ROCK CYCLE

Plutonism is igneous activity at depth in the crust, which produces the intrusive igneous rocks. In volcanism, ash and lava are erupted and extrusive igneous rocks are formed as the ejected material cools rapidly. These processes are associated primarily with three plate-tectonic settings: convergent boundaries, where oceanic plates descend into the mantle and melt and igneous rocks eventually form; divergent boundaries at mid-ocean ridges, where basaltic magmas rise from the mantle to form oceanic crust; and mantle plumes or hot spots in the interior of a continent, far from any plate boundaries, where magmas rise through the mantle and pour out at the surface.

On the continents, the circulation of the atmosphere weathers, transports, and deposits sediments by wind, water, and ice. Subsidence of an oceanic plate causes deposition, burial, and lithification of sediment transported from the continents.

Metamorphism and uplift occur where continental plates collide at convergent boundaries. These collisions uplift mountains and create the great pressures and temperatures that metamorphose rocks.

CHAPTER 3 OUTLINE

CLASSES OF IGNEOUS ROCKS

The two major classes of igneous rocks are intrusive and extrusive. Below is a list of characteristics and examples that are associated with one or the other type. On the lines below the two classes, write in the letters of the characteristics that best go with each class.

INTRUSIVE: EXTRUSIVE:

_____   _____

a. Large crystals   d. Formed in the interior   g. Slow cooling
b. Quick cooling   e. Granite   h. Formed on the exterior
c. Basalt   f. Tiny crystals   i. Glassy

PRACTICE MULTIPLE-CHOICE QUESTIONS

Circle the option that best answers the question.

1. The appearance of a rock depends on its:
   (a) Mineralogy and texture
   (b) Predominant mineral
   (c) Environment of formation
   (d) Degree of weathering

2. The two most abundant elements in most igneous rocks are:
   (a) Oxygen and silicon
   (b) Oxygen and aluminum
   (c) Silicon and iron
   (d) Iron and magnesium

3. Magma is the direct source material for which rock type?
   (a) Igneous
   (b) Sedimentary
   (c) Metamorphic
   (d) All three major rock types

4. The two major types of sediments and sedimentary rocks are:
   (a) Silicate and carbonate
   (b) Clastic and chemical/biochemical
   (c) Particles and ions
   (d) Plutonic and volcanic

5. The process that transforms a pile of loose sediment into a sedimentary rock is called:
    (a) Crystallization
    (b) Recrystallization
    (c) Lithification
    (d) Burial

6. Metamorphic rocks are those that have:
    (a) Been formed at Earth's surface
    (b) Undergone a great deal of erosion
    (c) Been heated to the melting point of their constituent minerals
    (d) Been changed by high temperatures and pressures

7. Regional metamorphism occurs when:
    (a) Only one localized region is affected by metamorphic processes
    (b) High pressures and temperatures affect large regions
    (c) The region around an igneous intrusion is heated up
    (d) Silicate minerals are selectively altered

8. Contact metamorphism most often occurs:
    (a) Over very large areas
    (b) During the process of mountain building when plates collide
    (c) When sedimentary minerals come into contact with metamorphic minerals
    (d) Along the borders of igneous intrusions

9. The processes of weathering and erosion produce which type of rock?
    (a) Igneous
    (b) Sedimentary
    (c) Metamorphic
    (d) All three major rock types

10. The rock cycle can be most accurately described by which order of events?
    (a) Plutonism, uplift, weathering and erosion, burial and lithification, metamorphism
    (b) Metamorphism, weathering and erosion, plutonism, uplift
    (c) Uplift, metamorphism, burial and lithification, plutonism
    (d) Weathering and erosion, plutonism, metamorphism, uplift

11. Uplift is most likely to occur in which of the following plate-tectonic settings?
    (a) Divergent boundary at a mid-ocean ridge
    (b) Subsidence of an oceanic plate
    (c) Hot spot or mantle plume in the interior of a continent
    (d) Convergent boundary where continents collide

CD RESOURCES

SUPPLEMENTAL DIAGRAMS

Figure 3.12   The rock cycle

SUPPLEMENTAL PHOTOS

See the supplemental photos for Chapters 4, 5, 7, and 8.

FIRST EDITION SLIDE SET

See the slides for Chapters 4, 5, 7, and 8.

EXPANSION MODULES

There are excellent photos of outcrops of all three major rock types in the expansion modules on Canada and regions of the United States (Chapters 25–30).

ILLUSTRATED GLOSSARY

EXERCISES

The Rock Cycle

TOOLS

The Rock Cycle

CD LAB

1. Magmas form in all plate-tectonic settings except:
    - (a) Subduction zones
    - (b) Continental collisions
    - (c) Transform boundaries
    - (d) Mid-ocean ridges

    (Hint: See the Rock Cycle tool.)

2. Which rock type is subject to burial and diagenesis?
    - (a) Igneous
    - (b) Sedimentary
    - (c) Metamorphic

    (Hint: See the Rock Cycle tool.)

3. Heat and pressure produce which type of rock?
    (a) Igneous
    (b) Sedimentary
    (c) Metamorphic

(Hint: See the Rock Cycle tool.)

Did you know that *Understanding Earth* has its own home page on the World Wide Web? The address is:

http://www.whfreeman.com/understandingearth

The page includes links to dozens of interesting sites you can visit for more information about many aspects of geology, organized by topic and by textbook chapter. If you're familiar with the Internet, fire up your web browser and check it out! If you need a little help getting started, click "Geology and the Internet" on the main menu of the CD-ROM.

To pique your interest, we've included a few "web projects" in the CD Lab sections of some of the chapters in this *Study Guide*. Here's the first:

Follow the instructions in "Geology and the Internet" or just go directly to the *Understanding Earth* home page. Click the Chapter 3 link and then the "All You Ever Need to Know about the Rock Cycle" link. Does the diagram look familiar? What do you think of it? Is it easier or harder to understand than the diagram in the textbook (Figure 3.10)? Try following some other links to see where they lead.

FINAL REVIEW

After reading this chapter, you should:

• Begin to understand how the texture and mineralogy of a rock are clues to its geological origins
• Be able to name and describe the three major rock types and how each is formed
• Begin to understand the differences in the chemical compositions of rocks
• Thoroughly understand the rock cycle
• Understand how the plate-tectonic settings where rocks form relate to the rock cycle

ANSWER KEY

CLASSES OF IGNEOUS ROCKS

Intrusive: a, d, e, g   Extrusive: b, c, f, h, i

PRACTICE MULTIPLE-CHOICE QUESTIONS

1. a, 2. d, 3. a, 4. b, 5. c, 6. d, 7. b, 8. d, 9. b, 10. a, 11. d

Igneous Rocks: Solids from Melts

This chapter provides a detailed view of the wide range of igneous rocks, both intrusive and extrusive, and the processes by which they form. It answers such questions as: How do igneous rocks differ from one another? Where do they form? How do rocks solidify from a melt? Where do melts form?

◆ Igneous rocks are classified on the basis of their texture and their mineral and chemical composition. Texture—whether a rock is fine-grained or coarse-grained—is linked to the speed and location of cooling. Slow cooling of magma in the interior allows time for the interlocking large, coarse crystals that characterize intrusive igneous rocks to grow. Rapid cooling at the surface produces the fine-grained texture or glassy appearance of the extrusive igneous rocks. A porphyry has a mixed texture in which large crystals "float" in a predominantly fine crystalline matrix The large crystals, called phenocrysts, formed while the magma was still below the surface. A volcanic eruption then brought the magma to the surface, where it quickly cooled to a finely crystalline mass.

◆ Modern classifications by chemical and mineral composition group igneous rocks by their relative proportions of silicate minerals (Table 4.1). As the mineral and chemical compositions of igneous rocks became known, geologists soon noticed that some extrusive and intrusive rocks were identical in composition and differed only in texture. Basalt, for example, is an extrusive rock formed from lava. Gabbro has exactly the same mineral composition as basalt but is formed deep in the crust. Extrusive and intrusive rocks form two chemically and mineralogically parallel sets of igneous rocks. *(See the accompanying article on the classification of igneous rocks.)*

CLASSIFICATION OF IGNEOUS ROCKS

Igneous rocks are classified by their chemical and mineral compositions and by their texture. Figure 4.6 shows the relationship between composition and texture for the nine major igneous rocks. An intrusive igneous rock and an extrusive igneous rock can have the same chemical composition but different textures; for example, rhyolite (a fine-grained extrusive rock) has the same chemical composition as granite (a coarse-grained intrusive rock). In this somewhat simplified classification scheme, the compositions of these rocks—as well as those of granodiorite/dacite, diorite/andesite, gabbro/ basalt, and peridotite—can vary within a certain range.

You can use Figure 4.6 to determine the mineral composition of a rock if you know how much silica it contains. Draw a vertical line perpendicular to the X axis at the appropriate silica content. This vertical line intersects the boundaries of the minerals found in this particular rock. Drawing a horizontal line from the Y axis to each point where the vertical line intersects a mineral boundary allows you to determine the approximate percentage of each mineral contained in the rock.

♦   We know that a rock can contain different minerals, each of which has its own melting temperature. When a rock is heated, each mineral melts as its particular melting temperature is reached. The entire rock does not melt all at once, but gradually. This process is called partial melting. As the amount of quartz in a rock increases, its melting temperature drops. Mafic magmas, such as those that form basalts, are hotter than the silicic magmas that produce granite. As mafic magmas rise, they often contain enough heat to melt more silicic rocks. Also, rising magma encounters rocks that are at shallow depths in the crust. The pressure in the shallower areas is lower, and rocks there will melt at lower temperatures. The presence of water is another factor that greatly lowers the temperature at which rocks melt.

♦   Magma chambers are magma-filled cavities in the lithosphere that form as rising drops of melted rock push aside surrounding solid rock. These chambers, which can be as large as several cubic kilometers, form in various plate-tectonic settings. (See the accompanying article on igneous rocks and plate tectonics.)

♦   Magmatic differentiation is the process by which a uniform parent magma leads to rocks of varying compositions. Magmatic differentiation occurs because different minerals crystallize at different temperatures. The composition of a cooling magma changes as it is depleted of the chemical elements withdrawn to make the crystallized minerals. Magmatic differentiation is kind of a mirror image of partial melting.

IGNEOUS ROCKS AND PLATE TECTONICS

We know from laboratory experiments that different kinds of rocks melt at different temperatures and pressures. This information gives us clues about where melting is likely to take place. Sedimentary rocks, for example, melt at much lower temperatures than does basalt. So we would expect sedimentary rocks to melt at much shallower depths in the crust than basalt. The geometry of plate motions is the link that ties melting to rock composition and tectonic activity. Igneous activity takes place in three main plate-tectonic settings: mid-ocean ridges, subduction zones, and mid-plate hot spots. Figure 4.8 is an excellent summary of plate-tectonic settings as they relate to magma composition.

At mid-ocean ridges, rising convection currents in the mantle form basaltic magmas. Basaltic magma forms in the hot upper mantle below mid-ocean ridges and rises to collect in shallow magma chambers near the crest of the ridge. Extremely large quantities of magma flow out of the rifts and fissures at mid-ocean ridges.

The magmas of subduction zones, formed partly from a mixture of seafloor sediments and partly from basaltic and felsic crust, can be basaltic or felsic. Water is carried downward with the seafloor sediments on the subducting plate. Because the presence of water greatly lowers the temperature at which rocks melt, at least partial melting occurs at the top of the subducting plate. The compositions of the materials that melt to become part of the magma determine the types of igneous rocks that form from the magma. Igneous rocks formed in subduction zones tend to be more silicic than the basalts of mid-ocean ridges. They include andesite and smaller amounts of more felsic volcanic rocks.

At hot spots far from plate boundaries, rising mantle plumes can cause the outpouring of huge quantities of basalt. Mid-plate mantle plumes can also cause the eruption of large quantities of basalt in isolated volcanic islands, such as the Hawaiian Islands.

Two sets of chemical reactions are explained by magmatic differentiation. In the continuous reaction series of plagioclase feldspars, there is a continuous, gradual series of chemical reactions. Calcium-rich plagioclase feldspar has the highest melting temperature. As this mineral forms, the remaining melt is depleted of calcium, thus making it richer in sodium. As the remaining melt crystallizes, the minerals that form are sodium rich. The key to this process is the continuous reaction of crystals with the melt. Each continuously changes by small amounts, so that at any point in the course of crystallization all crystals have the same composition.

BOWEN'S REACTION SERIES

N. L. Bowen studied how and when minerals crystallize as a basaltic magma cools. His reaction series (Figure 4.14) explains how the mineral assemblages found in many places could have formed. His scheme includes both the continuous reaction series of plagioclase feldspars and the discontinuous reaction series of mafic minerals such as olivine and pyroxene. In a continuous reaction series, the composition of the feldspars changes continuously and gradually as the magma cools. In a discontinuous reaction series, the cooling crystals change abruptly from one mineral to another at a particular temperature.

Bowen's reaction series was the first model to explain how high-silica rocks such as granite could form from an originally basaltic (low-silica) magma. Although his original theory of magmatic differentiation did not account for the large amounts of granite and granodiorite found on the continents and has since been supplanted by more complex theories, his work laid the foundation for most later investigations of differentiation in igneous rocks.

In the discontinuous reaction series, the composition of the crystals changes discontinuously as the magma cools, one mineral abruptly changing to another at a particular temperature. This pattern is characteristic of the mafic minerals such as olivine and pyroxene. The first mafic mineral produced is olivine. As the melt temperature decreases, the production of olivine is abruptly halted and a new mineral, pyroxene, is formed. There is no chemical association or continuity between the chemistry or appearance of the two minerals.

♦   Fractional crystallization is a process proposed by N. L. Bowen, a Canadian geologist, to account for the preservation of minerals formed earlier as the composition of the melt changed. The basic idea is that the first crystals to form could be segregated from the remaining melt, either by settling to the floor of the magma chamber or by structural deformation that would segregate and compress the crystals as a distinct intrusive body. Either way, crystals that had formed early would be segregated from the remaining melt, which would then behave as though it had just begun to crystallize. *(See the accompanying article on Bowen's reaction series.)*

♦   Plutons are large igneous bodies formed at depth in the crust. We can study plutons when uplift and erosion uncover them or when mines or drill holes cut into them. Plutons are highly variable in size, shape, and their relationship to the surrounding country rock. Batholiths, the largest plutons, are great irregular masses of coarse-grained igneous rock that cover at least 100 km^2. Similar but smaller plutons are called stocks. Both batholiths and stocks are discordant intrusions; that is, they cut across the layers of the country rock they intrude.

◆   Sills and dikes are similar to plutons, but they are smaller and have a different relationship to the layers of surrounding intruded rock. A sill is a tabular sheetlike body formed by injection of magma between parallel layers of preexisting bedded rock. Sills are concordant intrusions—that is, their boundaries lie parallel to these layers, whether or not the layers are horizontal.

Dikes are like sills in being tabular igneous bodies, but they cut across layers of bedding in country rock, rather than running parallel to them. Thus, they are discordant intrusions.

Veins are deposits of foreign minerals within a rock fracture. Pegmatites are veins of extremely coarse-grained granite cutting across a much finer-grained country rock.

CHAPTER 4 OUTLINE

CLASSIFICATION OF IGNEOUS ROCKS

For each of the 10 rock types listed on the left, select a corresponding term from each of the two lists on the right.

1. Granite   ____ ____
2. Basalt   ____ ____
3. Diorite   ____ ____
4. Rhyolite   ____ ____
5. Pumice   ____ ____
6. Granodiorite   ____ ____
7. Gabbro   ____ ____
8. Andesite   ____ ____
9. Obsidian   ____ ____
10. Dacite   ____ ____

I.   Felsic
II. Mafic
III. Intermediate
IV. Not applicable

a. Fine-grained
b. Coarse-grained
c. Glassy

PRACTICE MULTIPLE-CHOICE QUESTIONS

Circle the option that best answers the question.

1. A fine-grained texture is usually formed:
    (a) By a period of rapid cooling followed by a period of slow cooling
    (b) By a period of slow cooling followed by a period of rapid cooling
    (c) By relatively rapid cooling
    (d) When magma first crystallizes mafic minerals and then felsic minerals

2. Basalt and gabbro have the same _____ but different _____:
    (a) Textures, compositions
    (b) Compositions, textures
    (c) Cooling history, temperatures of formation
    (d) Amount of orthoclase, amount of ferromagnesium

3. The glassy texture observed in obsidian is a result of:
    (a) The presence of only one crystal in the rock
    (b) Cooling over a long period of time, during which the atoms actually break down and lose any order they had when the rock first formed
    (c) Alternating cooling and heating of the rock in the subsurface
    (d) Superrapid cooling such as that experienced when magma is ejected from a volcano

4. The silicate mineral on the discontinuous side of Bowen's reaction series having the most complex structure is:
    (a) Biotite
    (b) Pyroxene
    (c) Amphibole
    (d) Olivine

5. Intrusive igneous rocks are characterized by:
    (a) Fine-grained textures of predominantly mafic composition
    (b) Fine-grained textures of predominantly felsic or mafic composition
    (c) Coarse-grained textures of predominantly mafic composition
    (d) Coarse-grained textures of predominantly felsic or mafic composition

6. Which of the following pairs of minerals would not be found together in an igneous rock:
    (a) Olivine and pyroxene
    (b) Muscovite and quartz
    (c) Quartz and olivine
    (d) Amphibole and biotite

7. Magma bodies tend to rise in the crust because they:
    (a) Are large and need space to grow
    (b) Contain less dense minerals
    (c) Are hot, which has lowered their density
    (d) Are being forced up by the subduction process

8. Zoned crystals are a product of:
    (a) Fractional crystallization
    (b) Magmatic differentiation
    (c) Active subduction
    (d) Magmatic rise in the crust

9. An example of a tabular, discordant pluton is a:
    (a) Batholith
    (b) Stock
    (c) Sill
    (d) Dike

10. Which of the following characteristics would allow one to distinguish a sill from a lava flow if each is exposed at Earth's surface:
    (a) Their chemical compositions (that is, the minerals present) would be very different
    (b) Sills tend to be finer-grained, lava flows to be coarser-grained
    (c) Lava flows are usually smaller
    (d) Sills tend to be coarser grained than lava flows because they undergo slower cooling

11. Active plate subduction will produce:
    (a) Batholiths and andesitic volcanoes
    (b) Sills and dikes that are parallel to plate motion
    (c) No melting of any rocks near the subduction zone
    (d) Only minerals on the discontinuous side of Bowen's reaction series

CD RESOURCES

SUPPLEMENTAL DIAGRAMS

4.22   Types of igneous rock
4.23   Igneous rock types classified by sodium plagioclase and calcium plagioclase content
4.24   Successive stages in crystallization of a molten plagioclase feldspar
4.25   Zoned plagioclase crystal

SUPPLEMENTAL PHOTOS

4.26   Granodiorite
4.27   Diorite
4.28   Dacite
4.29   Dacite (top specimens) and andesite (bottom specimens)
4.30   Pegmatite dike in gneiss

FIRST EDITION SLIDE SET

Slide 4.1   Photomicrograph of gabbro
Slide 4.2   Photomicrograph of diabase
Slide 4.3   Mafic dike
Slide 4.4   Photomicrograph of basalt
Slide 4.5   Catalina granite
Slide 4.6   Sierra Nevada batholith, Yosemite National Park
Slide 4.7   Felsic dikes intruding Vishnu schist, Grand Canyon
Slide 4.8   Dike radiating from a volcanic neck

SECOND EDITION SLIDE SET

Slide 4.1   Photomicrograph of porphyritic andesite
Slide 4.2   Photomicrograph of zoned plagioclase phenocryst

ILLUSTRATED GLOSSARY

EXERCISES

Bowen's Reaction Series
Igneous Rock Properties

TOOLS

Igneous Rock Properties Databank

CD LAB

1. Which of the following is not an igneous rock?
    (a) Granite
    (b) Gabbro
    (c) Granulite
    (d) Granodiorite

    (Hint: See the Igneous Rock Properties databank.)

2. A coarse-grained igneous rock with intermediate silica content and abundant quartz is:
    (a) Diorite
    (b) Gabbro
    (c) Peridotite
    (d) Granodiorite

    (Hint: See the Igneous Rock Properties exercise.)

3. A zoned plagioclase crystal would have:
    (a) An albite-rich exterior
    (b) An anorthite-rich exterior
    (c) An albite-rich interior
    (d) Roughly equal amounts of albite and anorthite in interior and exterior

    (Hint: See the Chapter 4 supplementary diagrams.)

4. Which has the smaller crystals, gabbro or diabase?

    (Hint: See the first edition Slide Set.)

FINAL REVIEW

After reading this chapter, you should:

- Know the ways that igneous rocks are classified
- Know the nine major igneous rocks
- Be able to explain the formation of magma and the melting of rocks
- Understand what is meant by magmatic differentiation
- Understand the continuous and discontinuous reaction series
- Be able to explain Bowen's reaction series (Fig. 4.14, p. 92)
- Know the major forms of magmatic intrusions
- Be able to relate igneous rocks to plate tectonics

ANSWER KEY

CLASSIFICATION OF IGNEOUS ROCKS

1. I b, 2. II a, 3. III b, 4. I a, 5. IV c, 6. III b, 7. II b, 8. III a, 9. IV c, 10. III a

PRACTICE MULTIPLE-CHOICE QUESTIONS

1. c, 2. b, 3. d, 4. a, 5. d, 6. c, 7. c, 8. a, 9. d, 10. d, 11. a

5 Volcanism

About 80 percent of Earth's crust, oceanic and continental, is composed of volcanic rock. This chapter examines the process by which magma from Earth's interior rises through the crust, erupts onto the surface as lava, and cools into volcanic rock. The chapter covers the major types of lava, eruptive styles and the characteristic landforms they create, the relationship between volcanism and plate tectonics, the environmental disruption that volcanoes can cause and ways to mitigate that disruption, and some beneficial effects of volcanoes.

♦   Surface deposits of volcanic material are classified as either lava flows or pyroclastic deposits. Lava is what we call magma once it has erupted onto the surface. The major types of lavas and the rocks they form depend on the magmas from which they were derived.

Basaltic lavas erupt at very high temperatures and have low silica content, making them extremely fluid. Such lavas can spread quickly and widely. Basaltic lava flows vary according to the conditions under which they erupt. Flood basalts result when highly fluid basaltic lava erupts on flat terrain and spreads out in thin sheets. Cooling basaltic lava flowing downhill is classified as pahoehoe or aa. Pahoehoe forms when a highly fluid lava spreads in sheets and a thin, glassy elastic skin congeals on its surface as it cools. Aa is lava that has lost its gases and become more viscous than pahoehoe. It moves more slowly and forms a thick skin, which then breaks into rough, jagged blocks. Pillow lavas—piles of ellipsoidal, sacklike blocks of basalt about a meter wide—form in underwater eruptions.

Rhyolitic lavas erupt at lower temperatures than basaltic lavas, contain more silica, and are much more viscous. These lavas tend to pile up in thick, bulbous deposits. Andesitic lavas, with an intermediate silica content, have properties that fall between those of basalts and rhyolites.

♦   The presence of water and gases in magmas has a dramatic effect on eruptive styles. Before eruption, the confining pressure of the overlying rock keeps the water and gas from escaping. When the magma rises close to the surface and the pressure drops, the water and gas are often released with explosive force, shattering the lava and any overlying solidified rock into fragments of various sizes, shapes, and textures. Explosive eruptions are particularly likely with gas-rich, viscous rhyolitic and andesitic lavas.

♦   Pyroclasts are any fragmentary volcanic rock materials that are ejected into the air. When pyroclasts fall, they build deposits near their source. As they cool, the fragments become welded together (or lithified) into volcanic tuffs or volcanic breccias. A pyroclastic flow occurs when hot ash, dust, and gases are ejected in a glowing cloud that rolls downhill at speeds of up to 200 km per hour.

♦   The locations of volcanoes generally coincide with plate boundaries. *(See the accompanying article on volcanism and plate tectonics.)*

♦   Volcanic landforms depend on the properties of the lava and the conditions under which it erupts. Central eruptions discharge lava or pyroclastic materials from a central vent and create the familiar volcano shaped like a cone.

A shield volcano is a large, broad volcanic cone with very gentle slopes built up by successive flows of basaltic lavas. Mauna Loa on the island of Hawaii is a classic shield volcano. The eruption of felsic lavas produces a volcanic dome, a rounded, steep-sided mass of rock such as that formed by the 1980 eruptions of Mount St. Helens. A cinder-cone volcano results when a volcanic vent discharges pyroclasts and the solid fragments build up around the vent. Cerro Negro in Nicaragua is a cinder-cone volcano. When a volcano emits lava as well as pyroclasts, alternating lava flows and beds of pyroclasts build a concave-shaped composite volcano. This is the most common form of such large volcanoes as Fujiyama, Mount Vesuvius, and Mount St. Helens.

A bowl-shaped crater centered on the vent is found at the summit of most volcanoes. During an eruption, the upwelling lava overflows the crater walls. After the eruption, the lava that remains in the crater often sinks back into the vent and solidifies. With the next eruption, the material is blasted out of the crater, which then becomes partially filled by the debris that falls back into it.

After a violent eruption in which large amounts of magma are discharged from a magma chamber a few kilometers below the vent, the empty chamber may no longer be able to support its roof. The overlying volcanic structure can collapse, leaving a large, steep-walled, basin-shaped depression, much larger than the crater, called a caldera. Crater Lake, Oregon, fills a caldera 8 km in diameter.

VOLCANISM AND PLATE TECTONICS

The distribution of active volcanoes on Earth is one of the major pieces of evidence for plate tectonics. More than 400 active volcanoes are located near convergent plate boundaries. Most of the remaining 100 or so are found at divergent boundaries; a small percentage occur within plates, far from a plate boundary.

At the convergence of two ocean plates, an arc of volcanic islands builds up from the seafloor of the overriding plate. These volcanoes erupt mostly basaltic lavas. The Aleutian Islands are a classic example of this type of volcanism. At places where an oceanic plate converges with a continental plate, a volcanic mountain chain, such as the Cascade Range on the western coast of North America, develops in the collision zone near the continental margin. These volcanoes typically erupt large quantities of ash and andesitic lavas.

Eruptions that occur at divergent boundaries are usually hidden from view on the seafloor. Basaltic magma rises in the gap between the separating plates and overflows, creating mid-ocean ridges, volcanoes, and new seafloor crust. Iceland provides us with a rare first-hand look at what happens along a mid-ocean ridge, because this island is actually an exposed segment of the Mid-Atlantic Ridge. The eastern side of Iceland is part of the Eurasian Plate, and the western side is part of the North American Plate. Iceland is literally being pulled apart as the plates separate.

Scientists believe that volcanism occurring within plates is produced by fixed hot spots, or rising plumes of magma, in the mantle. The magma penetrates the lithosphere and erupts at the surface. Hot spots do not move with the lithospheric plates. As the plate passes over the hot spot, a string of extinct, progressively older volcanoes is left behind. The Hawaiian Islands were formed in this manner, and Yellowstone is also thought to be a result of hot spots.

When hot, gas-charged magma encounters groundwater or seawater, the vast quantities of superheated steam generated can cause a phreatic explosion. The 1883 eruption of Krakatoa in Indonesia was a phreatic explosion.

When hot matter from the deep interior escapes explosively, the vent and channel below can be left filled with breccia as the eruption wanes. The resulting structure is called a diatreme. Shiprock, New Mexico, is a diatreme exposed by the erosion of the sedimentary rocks through which it originally burst.

♦ A fissure eruption occurs when basaltic lava emanates from a long crack in Earth's surface rather than from a central vent. Some fissure eruptions occur along mid-ocean ridges; Iceland, which is an exposed segment of the Mid-Atlantic Ridge, experienced a catastrophic fissure eruption in 1783.

♦   Other volcanic features include lahars, torrential mudflows of wet volcanic debris; fumaroles, small vents through which volcanoes continue to emit fumes and steams for decades or centuries after an eruption; and geysers, hot-water fountains that spout intermittently with great force.

♦   Of Earth's 500 to 600 active volcanoes, one out of six has claimed human lives. Volcanoes kill people and damage property by edifice collapse, explosive blasts, ash falls, lethal gas release, lava flows, and mudflows or lahars. In many cases, geologists are now able to provide warnings of imminent major eruptions that allow people in the immediate vicinity to be evacuated. Volcanic eruptions cannot be prevented, but their catastrophic effects can be significantly reduced by a combination of science and enlightened public policy.

♦   Volcanoes also have some beneficial effects. Earth's atmosphere and oceans may have originated in volcanic episodes of the distant past. Soils derived from volcanic materials are exceptionally fertile because of the mineral nutrients they contain. Volcanic rock, gases, and steam are important sources of industrial materials and chemicals. Seawater circulating through fissures in mid-ocean ridges is a major factor in the formation of valuable ores. Thermal energy from volcanism is being harnessed; in Reykjavík, Iceland, for example, most of the houses are heated by hot water tapped from volcanic springs.

CHAPTER 5 OUTLINE

CENTRAL ERUPTIONS

Identify (1–3) the type of eruption and the type of volcano being shown in each figure below. In the space provided (4), write a paragraph explaining how differences in the ejecta of the three volcanoes cause their shapes to differ. What is it about the eruptive materials of each that makes the profiles different?

1. _____

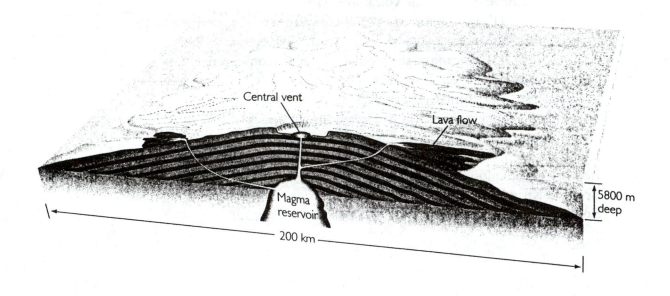

2. _____

3. _____

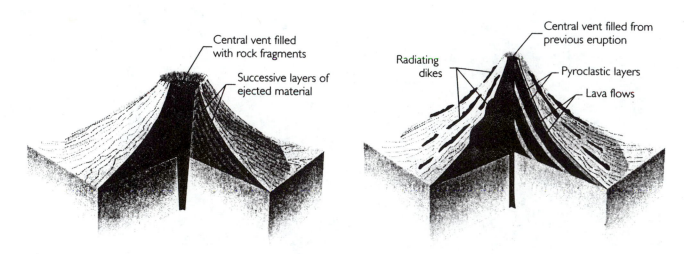

4. _____

_____

_____

_____

PRACTICE MULTIPLE-CHOICE QUESTIONS

Circle the option that best answers the question.

1. Fluid lavas have the following characteristics:
    (a) Low temperature and high silica content
    (b) High temperature and high silica content
    (c) Low temperature and low silica content
    (d) High temperature and low silica content

2. Pillow lavas are formed:
    (a) Underwater
    (b) By pyroclastic clouds
    (c) Close to the volcanic vent
    (d) At the front edge of a lava flow

3. Andesitic composite volcanoes would be most likely to form:
    (a) On a mid-ocean ridge
    (b) In the middle of a continent, away from all plate boundaries
    (c) Over a hot spot
    (d) Near an area of active plate subduction, such as along the west coast of South America

4. The violence of an eruption is related to:
    (a) The amount of gas contained in the lava
    (b) The location of the volcano (its closeness to a divergent or convergent boundary)
    (c) Whether it is a central eruption or a fissure eruption
    (d) The size of the fragments ejected

5. The formation of a volcanic dome is the result of:
    (a) The interaction of basalt and andesites
    (b) Highly viscous magmas, which are mainly felsic in composition
    (c) Exposing stocks that formed at shallow depths in the crust
    (d) Pyroclastic material accumulating on the surface near its source

6. A very large volcanic depression formed by the collapse of the land surface after an eruption is a:
    (a) Shield volcano
    (b) Composite volcano
    (c) Caldera
    (d) Cinder cone

7. A phreatic explosion is produced when:
    (a) Hot, gas-charged magma comes in contact with groundwater or seawater
    (b) Two lavas having different compositions mix with each other
    (c) Felsic lavas cool very rapidly
    (d) A volcanic crater collapses due to removal of the magma source

8. A lahar is:
    (a) The result of glacial ice coming into contact with lava from an eruption
    (b) A snow avalanche that results from eruptions at high elevations
    (c) A mudflow of wet volcanic debris
    (d) A fast-moving flow of basalt from a fissure eruption

9. Volcanic gases are important to the environment because:
    (a) They contain carbon dioxide
    (b) Their heat causes global warming
    (c) Inhaling the gases can kill people and animals
    (d) They can lower global temperatures and cause widespread crop failures

10. Geysers and hot springs are formed when:
    (a) Magma is extruded into water found on the surface
    (b) Circulating groundwater is heated by magma under the surface
    (c) Dissolved minerals are deposited by hot water
    (d) Large amounts of dissolved gases are released by magma

11. Iceland is an example of:
    (a) Plate subduction
    (b) Volcanism at a boundary between two converging oceanic plates
    (c) Volcanism at a boundary between two separating plates
    (d) Volcanism at boundary where an oceanic plate and a continental plate converge

12. The Hawaiian Islands formed because:
    (a) The Pacific Plate moved over a hot spot
    (b) Mid-oceanic basalts erupted and reached the surface of the water
    (c) Pyroclastic debris built up to reach sea level
    (d) A series of composite cones was produced by subduction

13. Which of the following is not a benefit of volcanism?
    (a) Geothermal energy
    (b) Fertile soils
    (c) Minerals and ores
    (d) Carbon dioxide released to the atmosphere

CD RESOURCES

SUPPLEMENTAL DIAGRAMS

5.34 Average composition of lava
5.35 Map of Hawaii showing locations of Kilauea and Mauna Loa
5.36 Cross section of an active Hawaiian volcano
5.37 Evolution of guyots from undersea volcanoes
5.38 Map showing Cascade Range volcanoes
5.39 Map of Mount St. Helens region
5.40 Cross sections of Mount St. Helens
5.41 Computer-graphic models of Mount St. Helens
5.42 Stages in the evolution of a resurgent caldera
5.43 Cross section of Long Valley Caldera
5.44 Krakatoa eruption cloud of 1883
5.45 Deep mantle plumes
5.46 Motion of Pacific Plate over fixed mantle plumes
5.47 Map of present world hot spots

SUPPLEMENTAL PHOTOS

5.48 Pahoehoe lava flows from the 1977 eruption of Krafla volcano, Iceland
5.49 Aa flow advancing through trees, Kilauea East Rift, Hawaii
5.50 Massive, blocky aa flow blocking road, Kilauea, Hawaii
5.51 Aa lava surge, Puu Oo eruption, Kilauea
5.52 Blocky aa lava flow surrounding church
5.53 Very hot, fluid lava erupting in a lava fountain.
5.54 Highly fluid lava erupting from fissures
5.55 Lava flows in three colors, Mauna Loa, Hawaii
5.56 Columnar jointed basalt flow, Washington
5.57 Volcanic ash from the eruption of Mount St. Helens
5.58 SEM of pumiceous ash in Denver from eruption of Mount St. Helens
5.59 Mauna Loa, a typical shield volcano
5.60 Arching lava fountain, Mauna Ulu, Kilauea
5.61 Lava fountains, Mauna Ulu, Kilauea
5.62 Lava fountain, Puu Oo eruption, Kilauea
5.63 Fountain and cascade of lava at night, Kilauea

FIRST EDITION SLIDE SET

Slide 5.1 Fountain eruption, Hawaii
Slide 5.2 Fissure eruption, Hawaii
Slide 5.3 Paricutin cinder cone eruption in 1944, Mexico
Slide 5.4 SP Crater, north of Flagstaff, Arizona
Slide 5.5 Lava tube, Kapaahu, east rift of Kilauea, Hawaii
Slide 5.6 Aa flow on top of pahoehoe, east rift of Kilauea, Hawaii
Slide 5.7 Flowing basaltic lava, east rift of Kilauea, Hawaii
Slide 5.8 Lherzolite nodule in basalt, San Bernardino volcanic field, southeastern Arizona
Slide 5.9 Olympus Mons volcano on Mars, *Viking Orbiter 1* image
Slide 5.10 Io, the innermost moon of Jupiter, *Voyager 1* image

SECOND EDITION SLIDE SET

Slide 5.1 Inyo Obsidian Domes on rim of Long Valley Caldera, Owens Valley, California
Slide 5.2 Pompeii, Italy
Slide 5.3 Cast of human victim buried by volcanic ash from the 79 A.D. eruption of Mount Vesuvius, Pompeii, Italy
Slide 5.4 Herculaneum, Italy
Slide 5.5 Anak Krakatau Volcano (the child of Krakatau), Sunda Strait between Java and Sumatra, Indonesia
Slide 5.6 Valley of the Ten Thousand Smokes, Katmai, Alaska
Slide 5.7 Thick layers of volcanic ash blanketing the slopes of Katmai Volcano, Alaska
Slide 5.8 Katmai Caldera, Alaska
Slide 5.9 Mount St. Helens, Washington
Slide 5.10 Popcatépetl Volcano, Mexico
Slide 5.11 Nevado del Ruiz Volcano, Colombia
Slide 5.12 The town of Armero, Colombia, devastated by a mudflow from Nevado del Ruiz

EXPANSION MODULES

Chapter 32: Volcanoes of the World (photo essay)

ANIMATIONS

5.1 Shield Volcano and Caldera Formation (Kilauea)
5.2 Composite Volcano (Mount Hood)
5.3 Cinder Cone Volcano (Northern Arizona)

ILLUSTRATED GLOSSARY

EXERCISES

Plumbing System of a Volcano
Volcanism and Plate Boundaries

TOOLS

Volcanoes of the United States and Canada

CD Lab

1. Which tectonic plate has the largest number of hot spots?
   (a) North American Plate
   (b) Pacific Plate
   (c) African Plate
   (d) Antarctic Plate

   (Hint: See the Chapter 5 supplemental diagrams.)

2. The release of what gas killed cattle at Lake Nyos, Cameroon, Africa, in 1986?
   (a) Sulfur dioxide
   (b) Carbon dioxide
   (c) Carbon monoxide
   (d) Nitrogen

   (Hint: See the Chapter 5 supplemental photos.)

3. Instead of silicate lava, volcanoes on Jupiter's moon Io erupt:

   (a) Hydrogen
   (b) Nitrogen
   (c) Sulfur
   (d) Carbon dioxide

   (Hint: See the first edition Slide Set.)

4. Katmai caldera in Alaska, formed during a giant eruption in 1912, is:
   (a) 2 kilometers across
   (b) 4 kilometers across
   (c) 6 kilometers across
   (d) 10 kilometers across

   (Hint: See the second edition Slide Set.)

5. An Indonesian volcano scarred by deep erosion gullies is:
   - (a) Batak volcano
   - (b) Merapi volcano
   - (c) Lamongan volcano
   - (d) Bromo volcano

   (Hint: See Expansion Module Chapter 32.)

6. Some of the eruptions that produced the Mono Craters in California occurred as recently as:
   - (a) 200 years ago
   - (b) 1000 years ago
   - (c) 2000 years ago
   - (d) 2500 years ago

   (Hint: See the Volcanoes of the United States and Canada tool.)

7. Garibaldi volcano in Canada last erupted:
   - (a) 200 years ago
   - (b) 1000 years ago
   - (c) 5000 years ago
   - (d) 10,000 years ago

   (Hint: See the Volcanoes of the United States and Canada tool.)

FINAL REVIEW

After reading this chapter, you should:

- Understand the relationship between types of lava and types of volcanic deposits
- Know the characteristics of the various eruptive styles
- Recognize the various volcanic structures and know their features
- Understand the relationship between volcanoes and plate tectonics
- Be able to explain the benefits as well as the limitations of human knowledge about volcanoes

ANSWER KEY

CENTRAL ERUPTIONS

1. Lava eruption; shield volcano
2. Pyroclastic eruption; cinder-cone volcano
3. Composite eruption; composite volcano or stratovolcano

4. Your answer should cover the following points: Basaltic lava flows easily and spreads widely, so in a lava eruption, the flows will create a broad, shield-shaped volcano. During pyroclastic eruptions, larger pyroclasts fall near the summit and form very steep, stable slopes. Smaller fragments slide downhill and form gentler slopes at the base. These eruptions result in the cinder-cone profile. Composite volcanoes are formed by alternating lava flows and eruptions of pyroclasts, giving them a profile in between those of the other two types.

PRACTICE MULTIPLE-CHOICE QUESTIONS

1. d, 2. a, 3. d, 4. a, 5. b, 6. c, 7. a, 8. c, 9. d, 10. b, 11. c, 12. a, 13. d

6

Weathering and Erosion

Weathering and erosion work continually on rocks exposed at Earth's surface. Weathering produces clays, soils, and dissolved substances that are carried by rivers to the ocean. Weathering can be caused by chemical reactions or by physical processes. Chemical weathering alters or dissolves the minerals in a rock. Physical weathering fragments a rock by physical processes that do not change its chemical composition. Chemical and physical weathering reinforce each other. The faster the chemical decay, the weaker the pieces of rock and the more susceptible they are to breakage; the smaller the pieces, the greater the surface area available for chemical attack and the faster the decay.

Erosion is the set of processes that loosen and move soil and rock downhill or downwind. Erosion moves weathered material on Earth's surface, carrying it away and depositing it elsewhere.

Weathering and erosion are major geological processes in the rock cycle. These processes change the form of Earth's surface and alter rock materials, converting igneous and other rocks into sediment and forming soil.

◆   All rocks weather, but how they weather and how fast they weather are controlled by four key factors: the properties of the parent rock, the climate, the presence or absence of soil, and the length of time the rocks are exposed to the atmosphere.

The properties of a parent rock affect weathering because different minerals weather at different rates and a rock's structure affects its susceptibility to cracking and fragmentation. Granite, for example, is resistant to weathering, whereas shale is easily weathered.

The rate of weathering is also controlled by climate. High temperatures and heavy rainfall speed chemical weathering; cold and dryness impede it. Physical weathering may be accelerated in a cold climate, where freezing water can widen cracks and push the rock apart.

The presence or absence of soil affects the chemical and physical weathering of rocks. Once soil starts to form, it works to weather rock more rapidly. Soil retains rainwater, and it hosts a variety of vegetation, bacteria, and organisms. These life-forms create an acidic environment that, in combination with moisture, promotes chemical weathering. The plant roots and organisms tunneling through the soil aid physical weathering.

The longer a rock is exposed, the more weathered it becomes.

♦   During chemical weathering, some minerals dissolve and others combine with water and components of the atmosphere, such as oxygen and carbon dioxide, to form new minerals. The text discusses the weathering of feldspar in detail because feldspars represent the most common group of silicate minerals. Their weathering has a major effect on the production of other minerals that make up sedimentary rocks. *(See the accompanying article on the weathering of feldspar.)*

♦   Clay minerals are a principal component of soils and sediments everywhere, because the rock-forming silicates constitute most of Earth's crust and weathering at the surface is widespread. Clays form through the weathering of a variety of silicate minerals in addition to feldspar. Amphibole, mica, and granite also weather to form clays. The chemical reactions by which these silicates weather to clay follow the same general course as feldspar weathering.

Not all silicates weather to form clay minerals. Rapidly weathering silicates, such as some pyroxenes and olivines, may dissolve completely in humid climates, leaving no clay residue. Quartz also dissolves without forming any clay mineral.

Silicate weathering can also form materials other than clay minerals. Bauxite—an ore composed of aluminum hydroxide—is one example.

♦   Iron silicates weather by the process of oxidation, which is the chemical reaction of an element with oxygen in which an ion or element loses one or more electrons. Oxidation is one of the important chemical weathering processes. Iron minerals, which are widespread, weather to the characteristic red and brown colors of oxidized iron.

♦   Limestone, made of the calcium and magnesium carbonate minerals calcite and dolomite, is one of the rocks that weather most quickly in humid climates. When limestone dissolves, no clay minerals are formed. The solid dissolves completely, and its components are carried off in solution. Much larger areas of Earth's surface are covered by silicate rocks than by carbonate rocks. But because carbonate minerals dissolve faster and in greater amounts than any silicates, the weathering of limestone accounts for more of the total chemical weathering of the land surface each year than that of any other rock.

CHEMICAL WEATHERING OF FELDSPAR

The text provides a detailed discussion of the weathering of granite, the primary rock present on the continents. Granite contains large amounts of potassium feldspar (orthoclase), along with quartz crystals. The weathering of feldspar by water produces a new mineral, kaolinite, along with dissolved silica and dissolved potassium ions. Kaolinite is a clay mineral that absorbs water into its crystal structure (that is, it is a hydrous mineral).

The reaction of feldspar with pure water is very slow, but the reaction rate can be increased by an acidic environment. Such an environment is created in nature when carbon dioxide in the atmosphere combines with water to produce carbonic acid (a weak acid). The carbonic acid and water react with feldspar to produce kaolinite, dissolved silica, potassium ions, and bicarbonate ions. The bicarbonate ion complex is quite soluble and ends up being carried away by any excess water. Figure 6.6 in the text shows this reaction schematically.

From the balanced equation describing the weathering of potassium feldspar, we can see that both water and carbonic acid must be present for the reaction to occur. The conditions required for the reaction are present when rocks are located in the ground, as opposed to being exposed to the air. For this reason, feldspar-rich rock, such as granite, buried in the subsurface undergoes chemical weathering much more rapidly than a slab of granite exposed to air.

◆   The weathering rates of minerals cover a great range, from the rapid rates of carbonates to the slow rate of quartz. The varying rates at which minerals weather reflect their chemical stability in the presence of water at given surface temperatures. The chemical weathering of silicates in Bowen's reaction series occurs most rapidly in those minerals produced under weathering conditions most dissimilar to those on Earth's surface. Olivine and calcium plagioclase are formed under the highest temperature and pressure conditions and are therefore very much out of equilibrium when they are exposed to conditions on the surface. The result is that they break down much more rapidly than do potassium feldspar, quartz, and muscovite mica, which are located at the lower end of Bowen's reaction series. The result of chemically weathering igneous rocks containing all the minerals in Bowen's reaction series is that olivine and calcium plagioclase weather first and potassium feldspar, quartz, and muscovite mica weather last. Table 6.2 shows the relative stability of minerals associated with the three major rock types.

◆   We can see the working of physical weathering most clearly in arid regions, where chemical weathering is minimal. Weathered outcrops in arid regions are covered by a rubble of fragments ranging in size from individual mineral grains only a few millimeters in diameter to boulders more than a meter across. The differences in size reflect varying degrees of physical weathering and patterns of breakage of the parent rock. As physical weathering continues, the larger particles are cracked and broken into smaller ones.

The chemical weathering that promotes physical weathering is itself promoted by fragmentation, which opens channels where water and air can penetrate and react with minerals inside the rock. The breaking up of the rock into smaller pieces exposes more surface area to weathering and so speeds the chemical reactions. Physical weathering is not always dependent on chemical weathering. There are processes by which unweathered rock masses are broken up, such as the freezing of water in cracks.

Rocks can break for a variety of reasons, including stress along natural zones of weakness and biological and chemical activity. Rocks have natural zones of weakness along which they tend to crack. In sedimentary rocks such as sandstone and shale, for example, these zones are the bedding planes formed by the successive layers of solidified sediment. Massive rocks tend to crack along regular fractures at intervals of one to several meters called joints. These fractures form while rocks are still deeply buried in the crust. Through uplift and erosion, the rocks rise gradually to Earth's surface. There, freed from the weight of tons of overlying rock, the fractures open slightly. Once the fractures open a little, both chemical and physical weathering work to widen the crack.

One of the most efficient mechanisms for widening cracks is frost wedging—breakage resulting from the expansion of freezing water. As the water freezes, it exerts an outward force strong enough to wedge open the crack and split the rock. Other expansive forces that can split rocks are generated when minerals crystallize from solutions in rock fractures. A recurring idea among geologists who research weathering is that rocks can break as a result of the daily alternation of hot days and cold nights in a desert. Part of the breakage process may be weakening of the rock caused by its expansion in the heat and contraction during the cold.

Exfoliation is a physical weathering process in which large flat or curved sheets of rock are fractured and detached from an outcrop. Spheroidal weathering is also a cracking and splitting off of curved layers from a generally spherical boulder, but usually on a much smaller scale.

Rivers excavate bedrock valleys by beating on the bedrock of the channel with transported rocks and by hurling their own force against the bedrock at waterfalls and rapids. Rock masses are also broken up by the scouring and plucking action of glaciers. Ocean waves pounding on rocky shores with a force equal to hundreds of tons per square meter also fracture exposed bedrock.

◆   The end product of weathering is a vertical sequence of loose material that contains various minerals, clays, oxides, and unweathered bedrock. This sequence is called regolith, and soil represents its uppermost portion. Subsurface exposures in a trench or roadcut will display a series of levels or horizons, each having a different appearance and composition. The uppermost horizon, the A-horizon, often contains organic material along with a small amount of minerals. If the soil is located in a relatively moist area, rainfall that percolates through the A-horizon leaches out soluble minerals. These dissolved minerals move down into the subsurface and accumulate in the B-horizon. Organic matter does not migrate into the B-horizon unless it is carried there by burrowing animals. Underneath the B-horizon is a layer of slightly altered bedrock, along with some clay. This constitutes the C-horizon.

The thicknesses of the different horizons depend on the parent material from which the soil was derived and on the chemical weathering conditions. Areas with fairly heavy annual rainfall have soils classified as pedalfers. In the western United States, pedocals predominate. These soils are characterized by the presence of soluble minerals at shallow depths; a general lack of rainfall prevents the minerals from being dissolved and carried deeper. The organic content of pedocals is also quite low. Laterites represent the extreme case of chemical weathering. They are confined to tropical regions, where copious rainfall leaches all soluble minerals from the soil and where the high temperatures cause intense alteration of the minerals.

Recently there has been much interest in ancient soils that have been preserved as rocks in the geological record, some of them more than a billion years old. These paleosols are being studied as guides to ancient climates and to the amounts of carbon dioxide and oxygen in the atmosphere in former times.

CHAPTER 6 OUTLINE

SOIL PROFILES

Using the figure, identify soil layers 1, 2, and 3, and complete the exercise by circling the option (in parentheses) that best completes the description of each layer.

1. _____

Usually not more than *(1–2, 2–4)* meters thick. Its topmost layer is usually the *(lightest, darkest)*, containing the *(lowest, highest)* concentration of organic matter.

In a thick soil that has formed over a long period of time, the inorganic components of this top layer are mostly clay and *(soluble, insoluble)* minerals, because most *(soluble, insoluble)* minerals have been leached from this layer.

2. _____

Organic matter is *(sparse, dense)*. *(Soluble, insoluble)* minerals and *(carbonates, iron oxides)* have accumulated in small pods, lenses, and coatings.

3. _____

Is *(highly, slightly)* altered bedrock, broken and decayed, mixed with clay from *(chemical, physical)* weathering.

PRACTICE MULTIPLE-CHOICE QUESTIONS

Circle the option that best answers the question.

1. Which of the following rocks would be most susceptible to weathering in a warm, moist climate?
    (a) Granite
    (b) Basalt
    (c) Limestone
    (d) All would undergo similar amounts of weathering

2. Fine-grained minerals will experience more rapid chemical weathering than coarse-grained minerals because:
    (a) Chemical bonds are weaker in the smaller grains
    (b) Smaller grains pack more closely together, which allows water to move more readily across the grains
    (c) Relatively more surface area exists in smaller grains for solutions to attack
    (d) Smaller grains are mechanically less sable, so they tend to fracture more easily

3. Which of the following is *not* a factor in the rate of weathering?
    (a) Climate
    (b) Presence or absence of soil
    (c) Properties of the parent rock
    (d) Atmospheric pressure

4. Kaolinite forms as a result of:
    (a) The chemical weathering of feldspar minerals
    (b) Increased amounts of silica in the soil
    (c) The physical weathering of granite
    (d) Oxidation changing the number of electrons in an iron cation

5. Hydration refers to the process of:
    (a) Losing chemical components during weathering
    (b) The altering of feldspar in rock particles to kaolinite
    (c) Gaining water
    (d) Losing water

6. Iron silicates, such as pyroxenes, weather by the process of:
    (a) Oxidation
    (b) Hydrolysis (through carbonic acid)
    (c) Acidic replacement of ions
    (d) Cation substitution

7. Which of the following groups is ordered from most stable to least stable?
    (a) Iron oxide, quartz, biotite mica, calcite
    (b) Calcite, biotite mica, quartz, iron oxide
    (c) Biotite mica, quartz, iron oxide, calcite
    (d) Iron oxide, calcite, quartz, biotite mica

8. Which type of rock accounts for the largest amount of total chemical weathering of the land surface?
    (a) Basalt
    (b) Limestone
    (c) Granite
    (d) Andesite

9. Which of the following is *not* a form of physical weathering?
    (a) Frost wedging
    (b) Cracks widening by plant root growth
    (c) Exfoliation
    (d) Dissolving of calcite by carbonic acid

10. Physical weathering will be most intense in which climate?
    (a) Warm and moist, where daily temperature variations are small
    (b) Cold and moist, where daily temperature variations are large and span the freezing point
    (c) Warm and dry, where daily temperature variations are small
    (d) Cold and dry, where daily temperature variations are large and span the freezing point

11. The leaching of soluble minerals occurs in the:
    (a) A-horizon
    (b) B-horizon
    (c) C-horizon
    (d) The A, B, and C horizons

12. Soils that are commonly found in arid climates are:
    (a) Rich in iron and magnesium
    (b) Pedosols that occur in the upper 2 m of the surface
    (c) Very rich in aluminum and serve as a major source of that metal
    (d) Pedocals that are enriched in calcium

13. Pedalfers are characterized by:
    (a) Large amounts of calcium minerals in their A-horizon
    (b) Containing aluminum and iron oxides and hydroxides
    (c) Being leached of soluble minerals in their upper layers
    (d) Being rather poor soils in which to grow crops

CD RESOURCES

SUPPLEMENTAL DIAGRAMS

6.19   Major factors controlling weathering and erosion
6.20   Earth's chemical reaction system
6.21   Feldspar weathering on two kinds of topography
6.22   Potassium feldspar dissolution rate
6.23   Dissolution of calcite or carbonate minerals
6.24   Transformation of igneous rock by weathering

SUPPLEMENTAL PHOTOS

6.25   Weathering of easily erodible sedimentary rocks
6.26   Joints, natural fractures that occur in most rock structures
6.27   Columnar joints in basalt
6.28   Devil's Tower, Wyoming (aerial view)
6.29   Salt or clay-induced physical weathering of layerd boulder
6.30   Exfoliating granite, Sierra Nevada, California
6.31   Weathering of granite boulder in exfoliated shells
6.32   Hoodoos of sandstone and mudstone in Goblin Valley, Utah
6.33   Double arch
6.34   Frost Wedging
6.35   Laterite soil
6.36   Pedocal soil
6.37   Pedalfer soil
6.38   Crack in building caused by expansive soil, Colorado

FIRST EDITION SLIDE SET

Slide 6.1   Boulder Beach, Reykjanes Peninsula, Iceland
Slide 6.2   Evidence of frost action, Sprengisandur area, central Iceland
Slide 6.3   Spearhead Mesa, Monument Valley, northeastern Arizona
Slide 6.4   Saltworks, Salt Lake, Utah
Slide 6.5   Red soil, Madera Canyon alluvial fan, southeastern Arizona
Slide 6.6   Paleosols in the Chinle formation, near Cameron, Arizona

SECOND EDITION SLIDE SET

See the slides listed for Chapters 11 and 12.

EXPANSION MODULES

The expansion module chapters on Canada and regions of the United States (Chapters 25–30) contain a number of photos showing examples of weathering and erosion. Chapter 37 (Historical Geology) shows some examples of differential weathering.

ILLUSTRATED GLOSSARY

EXERCISES

Mineral Stability and Weathering

CD LAB

1. Topography has an effect on how feldspar weathers. Would you expect to find arkose, a sandy sediment rich in feldspar, in an area of high mountains or low, rounded hills?
   (Hint: See Chapter 6 supplemental diagrams.)

2. Devil's Tower, Wyoming, is a good example of:
   (a) Joints
   (b) Paleosols
   (c) Feldspar weathering
   (d) Spheroidal weathering

   (Hint: See Chapter 6 supplemental photos.)

3. Rounded boulders are likely to be the result of erosion by:
   (a) Wind
   (b) Water
   (c) Ice
   (d) Chemical weathering

   (Hint: See the first edition Slide Set.)

4. Angular boulders are likely to result from:
   (a) Wind erosion
   (b) Water erosion
   (c) Chemical weathering
   (d) Physical weathering by frost or ice

   (Hint: See the first edition Slide Set.)

5. A hogback produced by differential weathering is likely to have:
   (a) An erosion-resistant sandstone layer on top, underlain by layers of weaker shale
   (b) A layer of weak shale on top, underlain by layers of erosion-resistant sandstone
   (Hint: See Expansion Module Chapter 37.)

6. Weathering and erosion are most likely to take place in which of the following plate-tectonic settings:
   (a) Divergent boundaries
   (b) Transform boundaries
   (c) Subduction zones
   (d) Stable continental interiors

   (Hint: See the Rock Cycle tool.)

FINAL REVIEW

After reading this chapter, you should:

- Know how rock type, climate, the presence of soil, and length of exposure control weathering
- Be able to describe how feldspar and other silicates chemically weather
- Know the relative stability of common minerals (see Table 6.2, page 147)
- Know the processes of physical weathering
- Understand how soil is formed through weathering
- Know the various soil profiles and soil groups

ANSWER KEY

SOIL PROFILES

1. A-horizon: 1–2, darkest, highest, insoluble, soluble
2. B-horizon: sparse, soluble, iron oxides
3. C-horizon: slightly, chemical

PRACTICE MULTIPLE-CHOICE QUESTIONS

1. c, 2. c, 3. d, 4. a, 5. c, 6. a, 7. a, 8. b, 9. d, 10. b, 11. a, 12. d, 13. b

Sediments and Sedimentary Rocks

Sedimentary rocks are records of the conditions at Earth's surface when and where the sediments were deposited, and so the study of sediments and sedimentary rocks is a key to understanding the current shape of Earth's surface and how it got that way. Sedimentary rocks are by far the most abundant rock type at the surface. Sediments and sedimentary rocks occur in the surface part of the rock cycle, between rocks brought up from the interior by tectonics and rocks returned to the interior by burial. The sedimentary stages of the rock cycle include weathering, erosion, transportation, deposition, burial, and diagenesis.

◆　　There are two basic kinds of sediments: clastic and chemical/biochemical. Clastic sediments are solid fragments produced by physical weathering of preexisting rocks. Chemical and biochemical sediments are dissolved substances precipitated from water by chemical and biochemical weathering. Clastic sediments are about 10 times more abundant in the crust than chemical and biochemical sediments.

◆　　After clastic sediments are formed by weathering, they are transported by currents of water or air and then deposited. Clastic sediments become sorted during transport as the strength and speed of the current vary. A strong, fast current may lay down a bed of gravel. When the current slows, a bed of sand is deposited on top of the gravel. When the current stops, a layer of mud is deposited on top of the sand bed. The tendency for variations in current velocity to segregate sediments according to size is called sorting. A well-sorted sediment contains particles of a uniform size. A poorly sorted sediment contains particles of many sizes.

◆　　Chemical and biochemical sediments dissolved during weathering are carried along with the water as a homogeneous solution and never settle out. As dissolved materials flow down rivers, they ultimately enter lake waters or the ocean. Such sediments are deposited by precipitation from the solution.

◆   A sedimentary environment is a geographic location characterized by a particular combination of geological processes and environmental conditions. Sedimentary environments are classified as continental (rivers, lakes, deserts, and glaciers); shoreline (deltas, tidal flats, and beaches); or marine (continental shelf, continental margins, organic reefs, and the deep sea). Clastic sedimentary environments include rivers, lakes, deserts, glaciers, deltas, beaches, the shallow continental shelf, and the deep sea. Chemical and biochemical sedimentary environments include carbonate, marine evaporite, deep-sea (siliceous), and swamp environments.

Different sedimentary environments exist at the same time in different parts of a region. A continental margin might include beach, tidal flat, and continental shelf environments, each with its characteristic kind of sediment. Such sets of sediments are called facies, and they describe the mineralogical, textural, and structural aspects of different sediments deposited simultaneously in the various environments.

◆   Structures formed at the time of deposition are called sedimentary structures. Bedding or stratification is a hallmark of sedimentary deposits. Cross-bedding, graded bedding, ripples, bioturbation structures, and bedding sequences are characteristic sedimentary structures. *(See the accompanying article on sedimentary structures.)*

◆   As sedimentation continues over time, thick piles of sediment may accumulate, partly in response to subsidence of the crust. Sedimentary basins are large regions where the combination of deposition and subsidence has formed thick accumulations of sediment and sedimentary rock. The accumulations found in these basins form geometrical shapes ranging from narrow troughs to circular or oval spoon-shaped depressions.

◆   After sediments are deposited and buried, they are subject to diagenesis. Diagenesis includes all changes in the properties of sediments or sedimentary rocks after they have been deposited, such as mineral composition and the amount of space between the grains. Burial promotes diagenesis because buried sediments are subjected to increasingly high temperatures and pressures in Earth's interior. Chemical changes include cementation and lithification. The major physical diagenetic change is compaction, a decrease in the volume and porosity of sediment.

◆   Clastic sediments and rocks are classified on the basis of their textures, primarily the sizes of the grains, into three broad categories: gravels and conglomerates (coarse); sand and sandstones (medium); and silt and siltstone; mud, mudstone, and shale; and clay and claystone (fine). Within each textural category, clastics can be further subdivided by mineralogy, which reflects the parent rocks.

SEDIMENTARY STRUCTURES

Sedimentary structures are strongly linked to the sedimentary environments in which they form. Cross-bedding in alluvial environments, for example, has a different geometry from that of cross-bedding found in desert dunes.

Cross-bedding consists of sets of bedded material inclined at angles up to 35° from the horizontal. Cross-beds form when grains are deposited on the steeper, downcurrent (lee) slopes of sand dunes on land or of sandbars in rivers and on the ocean floor. Cross-beds are also formed in sand dunes deposited by the wind.

In graded bedding, there is a progression from coarse grains at the base to fine grains at the top, reflecting a waning of the current that deposited the grains. A graded bed, which includes one set of coarse to fine beds, can range from a few centimeters to several meters thick.

Ripples are very small dunes of sand or silt whose long dimension is at right angles to the current. They form low, narrow ridges separated by wider troughs. Ripples form on the surfaces of windswept dunes, on underwater sandbars in shallow streams, and under the waves at beaches. Symmetrical ripples are made by waves moving back and forth on a beach. River currents or winds, which move in a single direction, form asymmetrical ripples.

Bioturbation structures are formed by marine organisms that excavate burrows and tunnels in and through beds of sediment. Bedding is broken or disrupted, sometimes crossed by small cylindrical tubes that extend vertically through several beds. From bioturbation structures, geologists can deduce the behaviors of organisms that burrowed the sediment and reconstruct the sedimentary environment.

Bedding sequences are patterns of interbedded sandstone, shale, and other types of sedimentary rocks. By examining these sequences, geologists can reconstruct how the sediments were deposited.

◆   Chemical and biochemical sediments and sedimentary rocks are classified by their chemical composition.

Carbonate sediments and sedimentary rocks are formed from carbonate minerals precipitated organically or inorganically during sedimentation or diagenesis. The minerals are either calcium carbonates or calcium-magnesium carbonates. Carbonate rocks are abundant because of the large amounts of calcium and carbonate present in seawater.

Evaporite sediments and sedimentary rocks are precipitated inorganically from evaporating seawater and from water in lakes in arid regions. Lakes that precipitate evaporites have no river outlet.

Silica sediments and sedimentary rocks are made up of chemically or biochemically precipitated silica. Much silica sediment is precipitated biochemically, secreted as shells by ocean-dwelling organisms.

Phosphorite is composed of calcium phosphate precipitated from phosphate-rich seawater. Phosphorite forms diagenetically by the interaction between muddy or carbonate sediments and the phosphate-rich water.

Iron formations usually contain more than 15 percent iron in the form of iron oxides and some iron silicates and iron carbonates. Most of these rocks formed early in Earth's history, when there was less oxygen in the atmosphere, and, as a result, iron was more easily soluble. In soluble form, iron was transported to the sea and precipitated there.

Coal is a biochemical sedimentary rock composed almost entirely of organic carbon formed by the diagenesis of swamp vegetation. Coal is classified as an organic sedimentary rock, a group that consists entirely or partly of organic carbon-rich deposits formed by the decay of once-living material that has been buried.

Oil and gas are fluids that are not normally classed with sedimentary rocks. But because they are formed by the diagenesis of organic material in the pores of sedimentary rocks, they can be considered organic sediments.

Chapter 7 Outline

CLASTIC SEDIMENT PARTICLE SIZES

Give the name of the missing clastic sediments and sedimentary rocks.

	SEDIMENT	PARTICLE SIZE	ROCK

COARSE

1

Boulder

—————— 256 mm ——————

2

—————— 64 mm ——————

3

5

—————— 2 mm ——————

Sand

6

—————— 0.062 mm ——————

Mud

4

—————— 0.0039 mm ——————

Siltstone

Clay

7
8
9

FINE

1. _____	4. _____	7. _____
2. _____	5. _____	8. _____
3. _____	6. _____	9. _____

PRACTICE MULTIPLE-CHOICE QUESTIONS

Circle the option that best answers the question.

1. Sediments composed of broken fragments of rocks and minerals are called:
    (a) Chemoclastic
    (b) Chemical
    (c) Clastic
    (d) Carbonates

2. The grain size of sandstone at a single outcrop can be used to infer:
    (a) The velocity of the current that deposited it
    (b) The distance over which the sand was transported
    (c) The number of times the sand was reworked by successive currents
    (d) The grain size of the rocks from which the sand was weathered

3. The sedimentary environment that would display the most poorly sorted group of clasts would be:
   (a) An alluvial fan
   (b) A deep-water marine area
   (c) An organic reef complex
   (d) A wind-deposited area

4. Sedimentary environments are satisfactorily characterized by:
   (a) The velocities of currents depositing the sediments
   (b) The mineralogical composition of the sediments
   (c) The location in which the sedimentation occurs
   (d) The temperature of the water in which the sediments are deposited

5. Cross-bedding is most prominent in:
   (a) Areas that experience rapid deposition
   (b) The quiet waters of deep ocean basins
   (c) Freshwater lakes
   (d) Wind or water depositional environments that have a prevailing direction of fluid flow

6. A vertical sequence of sedimentary clasts that displays a progression of large to small clast sizes from its bottom to its top is called:
   (a) A bedding sequence
   (b) A lithified layer
   (c) A coarse cross-bed
   (d) Graded bedding

7. A pattern of interbedded sandstone, shale, and other sedimentary rock types is:
   (a) A metamorphosed layer
   (b) Graded bedding
   (c) A bedding sequence
   (d) A bioturbation structure

8. The process by which unconsolidated sediment is transformed into solid rock is called:
   (a) Disintegration
   (b) Lithification
   (c) Agglomeration
   (d) Sorting

9. Clastic sediments are classified and named primarily on the basis of:
   (a) Particle size
   (b) Extent of lithification
   (c) Chemical composition
   (d) Type of source material

10. A clast whose average diameter is 6.5 mm would be classified as:
    (a) Silt
    (b) Clay
    (c) Pebble
    (d) Sand

11. Which sedimentary rock shows the widest range of particle sizes?
    (a) Sandstone
    (b) Conglomerate
    (c) Shale
    (d) Mudstone

12. Sedimentary rocks that contain very angular cobble-sized clasts are called:
    (a) Breccias
    (b) Conglomerates
    (c) Coarse shales
    (d) Coarse sandstones

13. The best-sorted sandstones, often found along beaches, are called:
    (a) Graywackes
    (b) Arkoses
    (c) Quartz arenites
    (d) Lithic arenites

14. Dolomite, the primary mineral in dolostones, is formed by:
    (a) Direct precipitation from fresh water
    (b) Foraminifera extracting minerals from seawater
    (c) The exposure of corals to sunlight
    (d) Diagenesis altering calcite

15. The rock that consists of fine-grained, chemically or biochemically precipitated silica is:
    (a) Sandstone
    (b) Gypsum
    (c) Halite
    (d) Chert

16. A biochemically produced sedimentary rock composed of organic carbon is:
    (a) Coal
    (b) Peat
    (c) Iron formation
    (d) Tar sand

CD RESOURCES

SUPPLEMENTAL DIAGRAMS

7.22 Downhill path of transportation and sedimentation
7.23 Ancient sedimentary environment in Guadalupe Mountains
7.24 Sequence of bed forms
7.25 Changes in bed forms
7.26 Typical marine deltaic cycle
7.27 Typical turbidity-current deposit
7.28 Classification of sedimentary rocks according to grain size

SUPPLEMENTAL PHOTOS

7.29 Marble Canyon in Grand Canyon
7.30 Living corals in shallow water
7.31 Phosphorite nodule in mudstone

FIRST EDITION SLIDE SET

Slide 7.1 Barnes conglomerate
Slide 7.2 Photomicrograph of a carbon cemented sandstone
Slide 7.3 Mudcracks, Death Valley, California
Slide 7.4 Fossil mudcracks
Slide 7.5 Death Valley from Dante's View

SECOND EDITION SLIDE SET

Slide 7.1 Photomicrograph of a foraminifer-rich micritic limestone
Slide 7.2 An atoll, a coral reef surrounding a central volcanic island

EXPANSION MODULES

Chapter 33: Sands of the World (Scientific American article)

ILLUSTRATED GLOSSARY

EXERCISES

Identifying Sedimentary Environments
Common Sedimentary Environments

TOOLS

Common Sedimentary Environments
Sedimentary Rock Properties databank

CD LAB

1. Which sequence of bedforms is one likely to find as a current increases in strength?
    (a) Sand waves, ripples, dunes, antidunes
    (b) Ripples, sand waves, dunes, antidunes
    (c) Ripples, dunes, antidunes, sand waves
    (d) Dunes, ripples, sand waves, antidunes
    (Hint: See the Chapter 7 supplemental diagrams.)

2. The Barnes Conglomerate is made up of:
    (a) Granite pebbles with a graywacke matrix
    (b) Granite pebbles with an arkosic sand matrix
    (c) Quartz pebbles with a graywacke matrix
    (d) Quartz pebbles with an arkosic sand matrix
    (Hint: See the first edition Slide Set.)

3. Mudcracks are a kind of:
    (a) Sedimentary structure
    (b) Sedimentary rock
    (c) Sediment
    (Hint: See the first edition Slide Set.)

4. Most beaches of northern Florida have sand composed of:
    (a) Quartz and calcereous materials
    (b) Colorless quartz
    (c) Quartz and feldspar
    (d) Feldspar and calcereous materials
    (Hint: See expansion module Chapter 33.)

5. Phosphorite is a sedimentary rock derived from which primary sediment?
    (a) Phosphorus sediment
    (b) Muds and clays
    (c) Calcereous sediment
    (d) None; it is formed by diagenesis
    (Hint: See the Sedimentary Rock Properties databank.)

6. A sedimentary environment found at the shoreline, where the primary transport agent is tidal currents and the primary sediments are sands and muds, is a:
    (a) Beach environment
    (b) Continental margin environment
    (c) Tidal flat environment
    (d) Continental shelf environment
    (Hint: See the Common Sedimentary Environments tool.)

Web Project: Take a virtual field trip to "Jurassic Reef Park" (click the link <u>A virtual field trip to the reefs of the Jurassic period</u> on the Chapter 7 page). As you explore the site, you might be thinking about questions like these: Why are reefs important? How were the ancient reefs of the Jurassic period different from modern reefs? How were they similar? What organisms built the ancient reefs? Are they the same as today's reef-building organisms? Were the ancient reefs very similar to one another, or were they diverse? What factors contributed to the pro-liferation of reefs in the Jurassic period?

FINAL REVIEW

After reading this chapter, you should:

- Understand the raw materials of sediment and how the raw materials relate to weathering
- Understand what happens during transportation and know the various methods of sediment transport
- Understand sedimentation and know the various sedimentary environments
- Know what diagenesis and lithification are
- Understand the text's discussion of bedding and sedimentary structures
- Understand how clastic and chemical/biochemical sediments are classified
- Know the major types of clastic and biochemical/chemical sediments and the rocks formed from them

ANSWER KEY

PARTICLE SIZE CLASSIFICATION OF CLASTIC SEDIMENTS

1. Gravel, 2. Cobble, 3. Pebble, 4. Silt, 5. Conglomerate, 6. Sandstone, 7. Mudstone, 8. Shale, 9. Claystone

PRACTICE MULTIPLE-CHOICE QUESTIONS

1. c, 2. a, 3. a, 4. c, 5. d, 6. d, 7. c, 8. b, 9. a, 10. c, 11. a, 12. c, 13. c 14. d, 15. d, 16. a

8 Metamorphic Rocks

Deep in the crust, temperatures and pressures are high enough to alter the mineral compositions, chemical compositions, and crystalline textures of sedimentary and igneous rocks, but not high enough to melt the rocks. This process of altera-tion is called metamorphism, and rocks that have undergone changes in one or more of these properties are classified as metamorphic rocks.

This chapter examines the causes of metamorphism, the types of metamorphism that occur under various conditions, and the origins of the textures that characterize metamorphic rocks.

♦ Most metamorphic rocks seen at the surface are produced by processes occurring at depths of 10 to 30 km. The pressure and heat that drive metamorphism result from three forces: Earth's internal heat, the weight of overlying rock, and the horizontal pressures that develop as rocks are deformed.

Temperature and pressure increase with depth. Metamorphic rocks formed at lower temperatures and pressures are classified as low-grade rocks, ones formed at higher temperatures and pressures as high-grade rocks. Sometimes a metamorphic rock is subjected early in its history to high pressures and tempera-tures, which produces a high-grade rock. Much later, the same rock encounters lower temperatures and pressures, which remetamorphose it to a low-grade rock. This process is called retrograde metamorphism.

♦ Heat affects a rock's mineralogy and texture by breaking chemical bonds and altering the existing crystal structures. The rock's atoms and ions recrystal-lize into new arrangements, creating new mineral assemblages. Many new crystals will grow larger than they were in the original rock, and the rock may become banded as different minerals are segregated into separate planes. Because each mineral crystallizes and remains stable at a characteristic temperature, we can use a rock's composition to gauge the temperature at which it formed.

◆   Pressure also changes a rock's texture and mineralogy. Solid rock is subjected to a general pressure in all directions that results from the weight of the overlying rock (confining pressure) and to pressure that is exerted in a particular direction, as when rocks are compressed and deformed by converging plates (directed pressure).

Under pressure, minerals may be compressed, elongated, or rotated to line up in a particular direction. For example, crystals of micas and amphiboles line up in planes perpendicular to the directed pressure. Because minerals recrystallize in characteristic ways at specific pressures, a rock's mineralogy and texture are indicators of the pressures under which it formed.

◆   The intrusion of a magma can cause chemical changes in surrounding rocks. Hydrothermal fluids rise from the magma, carrying various dissolved chemicals. These solutions can change the chemical and mineral compositions of a rock and sometimes replace one mineral by another without changing the rock's texture. This kind of change is called metasomatism.

◆   Although metamorphic rocks in outcrops appear to be completely dry and not porous, most do contain fluid in minute pores. Intergranular water containing dissolved carbon dioxide and other substances accelerates metamorphic chemical reactions. As metamorphism proceeds, both chemically bound water and pore water are lost from the rock. The higher the metamorphic grade, the lower the water content of the rock.

◆   Metamorphism can be categorized on the basis of the geological circumstances where it is found. *(See the accompanying article on metamorphism and plate tectonics.)*

Regional metamorphism occurs where both high temperature and high pressure are imposed over large belts of the crust. Some regional metamorphic belts are found near volcanic arcs that form where plates are subducted into the mantle. Others occur near oceanic trenches, where subduction drags down relatively cold oceanic crust. Regional metamorphism also occurs deep in the crust along converging continental plate boundaries.

Contact metamorphism occurs where magma intrudes into and changes the minerals of the surrounding rock. It is found along plate convergences, at oceanic and continental hot spots, and in deformed mountain belts.

Cataclastic metamorphism occurs primarily by the crushing and shearing of rock during tectonic movements. It produces the broken, pulverized texture found in strongly deformed mountain belts where faulting is extensive.

Hydrothermal metamorphism is often associated with spreading centers at mid-ocean ridges. Seawater heated by the hot, fractured basalts of the ridge flanks reacts chemically with the rock, forming altered basalts with changed chemical compositions. Hydrothermal metamorphism also occurs on continents when fluids rising from igneous intrusions metamorphose the penetrated rocks.

METAMORPHISM AND PLATE TECTONICS

Soon after the theory of plate tectonics was proposed, geologists started to see how patterns of metamorphism fit into the larger framework of plate-tectonic movements. Greenstones—metamorphosed basalts—are associated with metamorphism at mid-ocean ridges, where the seafloor spreads and basaltic magma wells up from the mantle. The heat of the magma transforms the newly extruded basalts into low-grade greenschists. The basalts are also altered by hydro-thermal circulation.

Blueschists, which are produced by very high pressures and relatively low temperatures, form in the forearc region of a subduction zone, the area between the seafloor trench and the volcanic arc. Sediments are carried down along the surface of the cool subducting plate. The plate moves down so quickly that it heats up very slowly, while the pressure in-creases rapidly.

High temperature, low pressure metamorphism also occurs at subduction zones, along the volcanic island arc on the overriding plate. The subducted plate melts, and the magma formed rises to shallow depths in the overriding plate. The hot magma transforms shallowly buried volcanics and sediments into greenschist and higher grade rocks. Thus, high pressure, low temperature metamorphism and low pressure, high temperature metamorphism are found as paired belts along plate convergences.

Regional metamorphic belts are associated with mountain building at continental collisions. In the cores of major mountain ranges, these belts of regionally metamorphosed rocks parallel the lines of folds and faults of the mountains.

Burial metamorphism is caused by the heat and pressure of overlying sediments and sedimentary rocks. Such metamorphic rocks are usually not strongly deformed but show broad, open folds. With the increasing temperatures and pressures that accompany orogenies, burial metamorphism grades into regional metamorphism.

♦ Metamorphism imprints new textures on the altered rocks. The texture of a metamorphic rock is determined by the size, shape, and arrangement of its crystals. Table 8.1 summarizes the textural classes of metamorphic rocks and their main characteristics.

The most prominent textural feature of regionally metamorphosed rocks is foliation, a set of flat or wavy parallel planes produced by deformation. Platy minerals tend to crystallize as thin platelike crystals whose planes are aligned parallel to the foliation. The parallel planes are called the preferred orientation of the minerals.

◆   The foliated rocks are classified according to four main criteria: the nature of their foliation; the size of their crystals; the degree to which their minerals are segregated into lighter and darker bands; and their metamorphic grade. The major foliated rocks are slate, phyllite, schist, and gneiss.

◆.   Nonfoliated rocks are composed mainly of crystals that grow in equant shapes, such as cubes and spheres, rather than in platy or elongate shapes. Nonfoliated rocks include hornfels, quartzite, marble, argillite, greenstone, amphibolite, and granulite.

◆   New metamorphic minerals may grow into large crystals, called porphyroblasts, surrounded by a much finer matrix of other minerals. Porphyroblasts are found in both contact and regionally metamorphosed rocks.

◆   Geologists seek to determine the intensity and character of metamorphism more precisely than is indicated by a designation of "low grade" or "high grade." There are various techniques available to make these finer distinctions.

Outcrops in a broad belt of regionally metamorphosed rocks show different sets of minerals. Various parts of these belts may be distinguished by their index minerals—the characteristic minerals that define metamorphic zones formed under a specific range of pressures and temperatures. Contact metamorphic zones are also characterized by index minerals that reflect different grades of metamorphism.

We can show on a map the zones where one metamorphic grade changes to another. Geologists define the zones by drawing lines called isograds, which connect the places where index minerals first appear.

The kind of rock that results from a given grade of metamorphism depends partly on the mineral composition of the parent rock. Metamorphic facies are groupings of rocks of various mineral compositions formed under different grades of metamorphism from different parent rocks.

CHAPTER 8 OUTLINE

TYPES OF METAMORPHIC ROCKS

The following figure is an incomplete representation of various types of meta-morphic rocks grouped in accordance with the temperature and pressure fields in which they were formed. Fill in the names of the missing groupings, or facies.

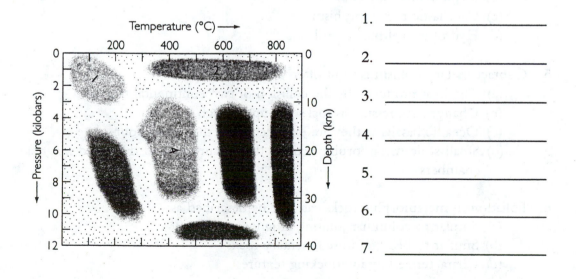

1. _____

2. _____

3. _____

4. _____

5. _____

6. _____

7. _____

PRACTICE MULTIPLE-CHOICE QUESTIONS

Circle the option that best answers the question.

1. As temperature and pressure increase in metamorphism, which of the following does *not* occur?
    (a) Minerals are aligned in preferred directions.
    (b) Water is driven off from the mineral structures.
    (c) Major changes in the bulk chemistry of the rocks take place.
    (d) Mineral grains increase in size.

2. Retrograde metamorphism refers to:
    (a) A high-grade rock being remetamorphosed to a low-grade rock
    (b) A low-grade rock being remetamorphosed to a high-grade rock
    (c) A high-grade rock being remetamorphosed to a higher grade rock
    (d) A low-grade rock being remetamorphosed to a lower grade rock

3. The alteration of a rock's bulk chemical composition by fluid transport into or out of the rock is called:
    (a) Hydrothermal metamorphism
    (b) Metasomatism
    (c) Chemical metamorphism
    (d) High-grade metamorphism

4. The deep burial of sediments in a large sedimentary basin would be an example of:
    (a) Burial metamorphism
    (b) Regional metamorphism
    (c) Cataclastic metamorphism
    (d) Hydrothermal metamorphism

5. Contact metamorphism is most often associated with:
    (a) The heat produced by the collision of two continental masses
    (b) Changes in pressure as rocks rise up through the crust
    (c) Oceanic basalts as they are extruded onto the ocean floor
    (d) Small-scale metamorphic activity that occurs around dikes and magma chambers

6. Foliation in metamorphic rocks can best be described as:
    (a) Similar to columnar jointing in igneous rocks
    (b) Similar to bedding in sedimentary rocks
    (c) Characterized by interlocking textures
    (d) A parallel alignment of the textural and structural features of a rock resulting from directed stress

7. Schistosity results from:
    (a) Growth of large, equidimensional grains
    (b) Alternating layers of light and dark minerals
    (c) Parallel alignment of large mineral grains
    (d) Total melting of all the minerals present

8. The progressive metamorphism of a shale will produce which of the follow-ing sequences of rock?
    (a) Schist, phyllite, slate
    (b) Migmatite, gneiss, granite
    (c) Slate, phyllite, schist
    (d) Claystone, sandstone, mudstone

9. An example of a nonfoliated metamorphic rock is:
    (a) Quartzite
    (b) Greenstone
    (c) Biotite schist
    (d) Migmatite

10. Which of the following terms is *not* associated with metamorphic processes?
    (a) Schist
    (b) Hornfels
    (c) Porphyroblast
    (d) Phenocryst

11. Mechanical deformation along a fault tends to produce:
    (a) Larger mineral grains
    (b) Fine-grained mylonites
    (c) Metasomatism
    (d) Decreased fluid flow

12. The metamorphic rock most likely to indicate that partial melting took place is a:
    (a) Shale
    (b) Slate
    (c) Schist
    (d) Migmatite

13. Greenstones are characteristic of metamorphism at:
    (a) A volcanic island arc
    (b) A mid-ocean ridge
    (c) A contact aureole
    (d) Forearc sediments in subduction zones

CD RESOURCES

SUPPLEMENTAL DIAGRAMS

8.18   Metamorphic zones in southern Vermont
8.19   Preferred orientation of elongate minerals
8.20   Schistosity

SUPPLEMENTAL PHOTOS

8.21   Photomicrograph of cataclastically deformed gneiss
8.22   Phyllite
8.23   Folded quartz vein cutting across schistosity in mica schist
8.24   Greenschist
8.25   Blueschist
8.26   Hornfels
8.27   Argillite
8.28   Greenstone
8.29   Amphibolite
8.30   Granulite
8.31   Mylonite
8.32   Migmatite

FIRST EDITION SLIDE SET

Slide 8.1   Metaconglomerate
Slide 8.2   Photomicrograph of quartzite
Slide 8.3   Photomicrograph of schist
Slide 8.4   Migmatite
Slide 8.5   Augen mylonitic gneiss

SECOND EDITION SLIDE SET

Slide 8.1   Laurel Mountain Roof Pendant, High Sierra, California
Slide 8.2   Roof pendant of metasediments in the Red Bluff Granite complex, Franklin Mountains, El Paso, Texas
Slide 8.3   Deformed Castner Marble in a roof pendant, Red Bluff Granite complex, Franklin Mountains, El Paso, Texas
Slide 8.4   Photomicrograph of a marble
Slide 8.5   Carrara marble quarry in the Apuan Alps, Italy

ILLUSTRATED GLOSSARY

EXERCISES

Metamorphism and Plate Tectonics

TOOLS

Metamorphic Rock Properties databank

CD LAB

1. In tightly folded bedded rocks:
    (a) Schistosity is perpendicular to the bedding planes
    (b) Schistosity is at an angle to the bedding planes
    (c) Schistosity is parallel to the bedding planes
    (d) There is no relationship between schistosity and the bedding planes
   (Hint: See the Chapter 8 supplemental diagrams.)

2. A metaconglomerate is:
    (a) A metamorphosed conglomerate
    (b) A conglomerate made up of quartz pebbles
    (c) A conglomerate that displays schistosity
    (d) A conglomerate that displays foliation
   (Hint: See the first edition Slide Set.)

3. A nonfoliated, low-grade rock whose principal mineral is chlorite is:
    (a) Argillite
    (b) Greenstone
    (c) Amphibolite
    (d) Granulite
   (Hint: See the Metamorphic Rock Properties databank.)

4. Slate is metamorphosed primarily from:
    (a) Shale
    (b) Sandstone
    (c) Basalt
    (d) Granite
   (Hint: See the Metamorphic Rock Properties databank.)

FINAL REVIEW

After reading this chapter, you should:

- Understand the physical and chemical factors controlling metamorphism and know the various types of metamorphism
- Know the metamorphic textural varieties and something about the metamorphic processes that created them

- Know something about how and why the texture and mineralogy of metamorphic rocks is studied
- Be familiar with the text's discussion of contact metamorphic zones
- Understand how metamorphism relates to plate tectonics

ANSWER KEY

TYPES OF METAMORPHIC ROCKS:

1. Zeolite, 2. Hornfels, 3. Blueschist, 4. Greenschist, 5. Amphibolite, 6. Granulite, 7. Eclogite

PRACTICE MULTIPLE-CHOICE QUESTIONS

1. c, 2. a, 3. b, 4. a, 5. d, 6. d, 7. c, 8. c, 9. a, 10. d, 11. b, 12. d, 13. b

The Rock Record and the Geologic Time Scale

The geological processes that shape Earth's surface and give structure to its interior work over millions and billions of years. This chapter examines how geologists study such slow processes and use the knowledge gained to reconstruct Earth's history. It tells the story of how, using a variety of dating techniques, geologists have been able to construct a geologic time scale of the entire history of the Earth.

◆   Geologists deal with Earth processes that take place over a great range of time periods, from seconds or minutes (earthquakes) to many millions of years (the building of a mountain chain). We can measure some geological processes directly, such as a river flooding or even a glacier advancing. But the only way we can time geological processes that are too slow to be measured directly is from the rock record. Rocks record geological events, such as glaciations, that lasted many thousands or millions of years.

◆   There are two kinds of age that interest geologists. Relative age is the age of sedimentary rocks layers in relation to one another. Radiometric (or "absolute") age is the actual number of years that have passed since a rock formed.

◆   In the nineteenth century, geologists built a geologic time scale from the time and space relationships of outcrops or of rocks exposed in drill holes. They did this using stratigraphy—the description, correlation, and classification of strata in sedimentary rocks. Stratification is basic to two principles used to interpret geological events from the sedimentary rock record. The principle of original horizontality states that sediments are originally deposited as essentially horizontal beds. The principle of superposition states that each layer of sedimentary rock in a tectonically undisturbed sequence is younger than the one beneath it and older than the one above it. Thus, a vertical set of layers, called a stratigraphic sequence, is a kind of time line—a partial or complete record of the time elapsed from the deposition of the lowest bed to the deposition of the uppermost bed.

Stratigraphy gauges relative age, not radiometric or "absolute" age. Sediments do not accumulate at a constant rate, nor does the rock record tell us how many years have passed between periods of deposition. Finally, stratigraphy alone cannot determine the relative ages of two widely separated beds.

◆   Fossils are the key to detecting missing time intervals and correlating the relative ages of rocks at different geographic locations. Fossilized remains of organisms are found in sedimentary rocks throughout the Earth. The discovery of fossils allowed geologists and paleontologists to surmise that life has existed for a very long period of time and that there has been a progressive development in its form and complexity. Charles Darwin used fossils to help formulate his theory of evolution. William Smith, an engineer and surveyor working in England, showed that the widespread occurrence of fossils could be used to correlate rock sequences in different locations. He noted that different layers contained different kinds of fossils, and he was able to tell one layer from another by the characteristic fossils in each. He established a general order for the sequence of fossils and strata, from lowermost (oldest) to uppermost (youngest) rock layers. This stratigraphic ordering of the fossils is known as a faunal succession.

Smith used faunal succession to correlate rocks from different outcrops. In each outcrop, he identified distinct formations. A formation is a series of rock layers that has about the same physical properties and contains the same assemblage of fossils. Some formations consist of a single rock type; others are made up of thin interbedded layers of different rock types. Each formation comprises a distinctive set of rock layers that can be recognized as a unit.

◆   In putting together sequences of formations, geologists often find places in which a formation is missing: either it was never deposited, or it was eroded away before the next strata were laid down. The boundary along which the two existing formations meet is called an unconformity. Like sedimentary rock layers, an unconformity represents time. Geologists classify different kinds of unconformities in terms of the relationships between the upper and lower sets of layers. An angular unconformity is one in which younger sediments rest atop the eroded surface of tilted or folded older rocks. In a disconformity, the upper set of layers overlies an erosional surface developed on an undeformed, still horizontal lower set of beds. A nonconformity is an unconformity in which the upper beds overlie metamorphic or igneous rocks.

♦   Cross-cutting relationships are also important clues for dating. Because we know that deformations or igneous intrusions that cut across sedimentary layers must have occurred after the sedimentary layers were deposited, they must be younger than the rocks they cut.

♦   Sequence stratigraphy is another, newer method of relative age dating. It is based on a sequence, which is a series of sedimentary beds bounded above and below by unconformities. These unconformities represent fluctuations in sea level, which allowed erosion to take place. The beds that make up the sequences show internal patterns that reveal changes in sedimentation. Sequence stratigraphers can match sequences of the same geologic age over wide areas and reconstruct a region's geologic history, including changes in sea level, in terms of successions of sequences.

♦   During the nineteenth and twentieth centuries, geologists used information from these relative dating principles to work out the geologic time scale, a relative-age calendar of Earth's geologic history. The geologic time scale is divided, in order of decreasing length, into four major time units: eons, eras, periods, and epochs. Each time interval on this scale is correlated with a corresponding set of rocks and fossils. Although the geologic time scale is still being refined here and there, its main divisions have remained constant for the past century.

♦   The geologic time scale based on studies of stratigraphy and fossils is a relative one; with it, geologists can say whether one formation is older than another, but they cannot pinpoint precisely when a rock formed. Radiometric dating, the use of naturally occurring radioactive elements to determine the ages of rocks, can provide an absolute age in numbers of years. *(See the accompanying article on radiometric dating.)*

Once geologists had determined radiometric ages and linked them to their earlier studies of fossils and stratigraphy, they could add absolute dates to the geologic time scale. After almost a century of radioactive dating and continued work on the stratigraphy of the world, this time scale remains undisputed in all major respects. *(See the Geologic Time Scale tool on the CD.)*

♦   The geologic time scale can tell us about the rates of some slow geological processes, such as mid-ocean spreading in the South Atlantic, the vertical movement in the uplift of the Alps, and the erosion of the North American continent. In these particular cases, it took about 100 million years to open an ocean, 15 million years to raise a mountain range, and 100 million years to erode it. Yet these time intervals are relatively short compared with the entire history of the planet. During that 4.6-billion-year history, Earth has experienced many cycles of mountain building and erosion.

RADIOMETRIC DATING

The discovery of radioactivity in certain elements was put to use by geologists to determine the age of rocks. Radiometric dating was of major importance because it allowed scientists, for the first time, to assign specific ages to rocks and minerals.

Before reading the text material on radioactive decay, you might want to reread the discussion of atomic structure in Chapter 2. More than 80 elements have radioactive isotopes, but only a few are useful for dating rocks. The most important of these are listed in Table 9.1.

The material that decays is called the parent isotope, and the products of the decay are called daughter isotopes. Because matter is neither created nor destroyed, the total mass of a decaying system remains constant. This fact, and the fact that the decay rate does not change, allows scientists to determine the ratio of parent to daughter isotopes.

To assign an age to a sample for which the ratio of parent atoms to daughter isotopes has been calculated, we need to know the half-life of the particular radioactive isotope. (The half-life is the time it takes for half of the parent atoms to decay to daughter isotopes.) The authors discuss the element rubidium (atomic weight 87) to illustrate the concept of natural radioactive decay. In the case they discuss, in which 950 parent atoms of rubidium-87 and 50 daughter atoms of strontium-87 are counted in a sample, 95 percent of the original atoms are left (950 + 50 = 1000 original atoms of Rb-87). From the known half-life of rubidium (47 billion years) and the remaining fraction of parent material, we can calculate that the age of this sample is about 4 billion years.

Carbon-14, with a half-life of 5730 years, is an important tool for dating organic materials such as bones, shells, and wood in very young sediments (100 to 70,000 years old). All organic matter contains carbon, including small amounts of carbon-14. Carbon-14 decays to nitrogen-14, and the age of a sample can be calculated by comparing the amount of carbon-14 remaining in the sample with the original amount that would have been in equilibrium with the atmosphere.

Radiometric dating does not always work. Problems arise when the amount of remaining parent material is very small or when the daughter isotopes have been removed from the sample, such as by weathering or metamorphism. Still, the precision of radiometric dating has improved greatly in recent years, and accurate ages can now be determined from extremely small samples.

CHAPTER 9 OUTLINE

RADIOACTIVE ISOTOPES

In the table below, fill in the numbered blanks to complete the table.

ISOTOPES	HALF-LIFE	MATERIALS THAT CAN BE DATED
Uranium–lead	1. _____	2. _____ Uraninite
Potassium–argon	1.3 billion years	3. _____ 4. _____ 5. _____ Whole volcanic rock
Rubidium–strontium	47 billion years	Muscovite Biotite 6. _____ Whole metamorphic or igneous rock

PRACTICE MULTIPLE-CHOICE QUESTIONS

1. The principle of original horizontality states that:
    (a) Sediments are deposited as essentially horizontal beds
    (b) All rocks form in a horizontal manner
    (c) Rocks that form horizontally always remain that way
    (d) Rocks that form horizontally never remain that way

2. Relative age can be determined by:
    (a) Noting which rocks are younger or older than surrounding rocks based on their stratigraphic relationships by using the principle of super-position
    (b) Determining which rocks have weathered the most intensely
    (c) Finding the amount of parent and daughter products present in a rock sample
    (d) Measuring the thicknesses of the rocks

3. A general order for the sequence of fossils and strata, from lowermost (oldest) to uppermost (youngest) rock layers is:
    (a) Sequence stratigraphy
    (b) A faunal succession
    (c) A stratigraphic sequence
    (d) A cross-cutting relationship

4. A series of rock layers that has about the same physical properties and contains the same assemblage of fossils is:
    (a) A stratum
    (b) A sedimentary bed
    (c) A sequence
    (d) A formation

5. If a stratigraphic sequence is missing a layer that should be present, the sequence is said to have:
    (a) An angular unconformity
    (b) A graded bed
    (c) An unconformity
    (d) A cross-cutting relationship

6. Given that during a half-life one-half of the parent atoms decay, what fraction of the parent material remains after four half-lives have passed?
    (a) 1/2
    (b) 1/4
    (c) 1/8
    (d) 1/16

7. The radioactive element that would be most appropriate for dating rocks that formed roughly 4 to 5 million years ago would be:
   - (a) Uranium-238
   - (b) Potassium-40
   - (c) Rubidium-87
   - (d) Carbon-14

8. Radioactive carbon-14 is not used to date old rocks because:
   - (a) It has a very low disintegration constant
   - (b) It has a very short half-life
   - (c) It is a very rare isotope and cannot be isolated
   - (d) Carbon-12 is present in much greater amounts

9. The Phanerozoic eon includes:
   - (a) Only time since 542 million years ago
   - (b) Only time before 542 million years ago
   - (c) The period between 2500 and 542 million years ago
   - (d) All geologic time

10. The time scheme that couples absolute and relative ages is the:
    - (a) Principle of superposition
    - (b) Principle of cross-cutting relationships
    - (c) Geologic time scale
    - (d) Radiometric dating

CD RESOURCES

SUPPLEMENTAL DIAGRAMS

9.18   Faunal succession in a lithologic sequence
9.19   Part of stratigraphic succession mapped by William Smith
9.20   Evolution of an unconformity
9.21   Evolution of an angular unconformity
9.22   Unconformity, disconformity, nonconformity
9.23   Relative time dating by field relations
9.24   Stratigraphic interpretation of Grand Canyon Bright Angel Shale
9.25   (A) Temporal span of Paleozoic units of the Grand Canyon shown in (B)
9.26   Geologic time
9.27   Geologic time units and chronostratigraphic units.
9.28   The emergence of life
9.29   The geologic cycle as deduced by Hutton
9.30   Age bracketing by radiometric dating of igneous rocks
9.31   Precambrian age map of Canada

SUPPLEMENTAL PHOTOS

9.32   Ancient and modern cephalopods
9.33   Stromatolitic fossil algal mounds, Montana
9.34   Dinosaur footprint in Dakota sandstone
9.35   Dinosaur tracks, northern Arizona
9.36   Fossil *Glossopteris* leaf, Permian age
9.37   Liquid scintillation radon detector
9.37   Alpha track radon detector

FIRST EDITION SLIDE SET

Slide 9.1   Trilobite (*Olenoides seradtus*), Burgess shale, Cambrian age
Slide 9.2   Fossil fern, Llewellyn formation, Pennsylvania
Slide 9.3   Fossil sycamore-like leaf, Green River Formation, Eocene
Slide 9.4   Angular unconformity, Shinumo Creek, Grand Canyon, Arizona
Slide 9.5   Grand Canyon from the Kaibab Trail, south rim, Arizona
Slide 9.6   Argon analysis by mass spectrometer for a potassium–argon radiometric age, University of Arizona, K–Ar Laboratory

SECOND EDITION SLIDE SET

See the slides listed for Chapters 7 and 24.

EXPANSION MODULES

Chapter 37: Historical Geology

ILLUSTRATED GLOSSARY

EXERCISES

Geologic Time Scale
Field Relationships for Relative Time Dating

TOOLS

Geologic Time Scale

ANIMATIONS

20.4   Plate Tectonics: Transform Boundary

CD LAB

1. Geologists use both time units and "chronostratigraphic" units. The former units measure periods of time; the latter, series of rock layers. What is the chronostratigraphic equivalent of the following time periods?
   (a) Late Devonian
   (b) Early Permian
   (c) Cretaceous

   (Hint: See the Chapter 9 supplemental diagrams.)

2. The interval of time between the first and last appearance of a particular species or group of fossils is called its:
   (a) Geologic range
   (b) Index period
   (c) Index range

   (Hint: See the first edition Slide Set.)

3. Flowering plants range from:
   (a) Triassic to Holocene
   (b) Jurassic to Holocene
   (c) Triassic to Pleistocene
   (d) Jurassic to Pleistocene

   (Hint: See the first edition Slide Set.)

4. The earliest bones of the genus *Homo* we have found so far are of:
   (a) Eocene age
   (b) Miocene age
   (c) Pliocene age
   (d) Pleistocene age

   (Hint: See the Geologic Time Scale tool.)

5. Which era is subdivided into the most periods?
   (a) Paleozoic
   (b) Mesozoic
   (c) Cenozoic

   (Hint: See the Geologic Time Scale tool.)

6. "Camels often sit down carefully. Perhaps their joints creak. Early oil might prevent permanent rheumatism." This mnemonic device (memory aid) is sometimes used to recall the order of the periods of the Paleozoic and Mesozoic eras and the epochs of the Cenozoic. Can you list the periods and epochs in their correct order by using this device?
   (Hint: See the Geologic Time Scale tool.)

7. "The present is the key to the past" is another way of describing the:
   (a) Principle of superposition
   (b) Principle of uniformitarianism
   (c) Principle of original horizontality
   (d) Principle of isostasy

(Hint: See Expansion Module Chapter 37.)

8. "Delicate Arch" in Arches National Park, Utah, was produced by:
   (a) Chemical weathering
   (b) Physical weathering
   (c) Differential weathering
   (d) Volcanism

(Hint: See Expansion Module Chapter 37.)

FINAL REVIEW

After reading this chapter, you should:

* Have an expanded idea of how geologists deal with the concept of time
* Understand how fossils help geologists to determine relative age
* Know how radiometric dating works
* Know how to order rock strata
* Know the names and sequences of epochs, periods, eras, and eons in the geologic time scale
* Understand why the geologic time scale is crucial to geology

ANSWER KEY

RADIOACTIVE ISOTOPES

1. 4.5 billion years, 2. Zircon, 3. Muscovite, 4. Biotite, 5. Hornblende, 6. Potassium feldspar

PRACTICE MULTIPLE-CHOICE QUESTIONS

1. a, 2. a, 3. b, 4. d, 5. c, 6. d, 7. b, 8. b, 9. a, 10. c

Folds, Faults, and Other Records of Rock Deformation

Deformation in Earth's crust is recorded as folding and faulting, the bending and breaking of rock. This chapter examines rock deformation and how geologists interpret field observations of deformation to reconstruct the geologic history of a region. From deformation patterns, geologists can tell what kind of rocks were originally deposited, how they were deformed, what kinds of forces were involved, and the plate-tectonic settings where the deformation occurred.

♦   A fold is a bent or warped rock layer or sequence of layers that was originally horizontal and was later deformed. A fault is a planar or gently curved fracture in the Earth's crust across which there has been displacement. Tectonic forces are ultimately responsible for the deformations found in most local areas.

♦   Geologists need accurate information about the geometry of visible outcrops to figure out how rock formations are deformed. The orientation of the observed layer is an important clue in piecing together a picture of the overall deformed structure. The orientation of an exposed rock layer can be described by two measurements: the strike and the dip. The strike is the direction taken by a rock layer as it intersects the horizontal. The dip, which is measured at right angles to the strike, is simply the amount of tilting, or the angle at which the bed inclines from the horizontal.

Geologists use geological maps to record the locations of outcrops, the nature of their rocks, and the dips and strikes of inclined layers. They also use geological cross sections, diagrams showing the features that would be visible if a vertical slice were made through part of the crust.

◆   There are three types of tectonic forces that produce folding and faulting: compressive forces squeeze and shorten a body; tensional forces stretch a body and tend to pull it apart; and shearing forces push two sides of a body in opposite directions. These forces are associated with the three types of plate boundaries. *(See the accompanying article on deformation and plate tectonics.)*

    Laboratory experiments in which different types of rocks are squeezed at various temperatures and pressures have been conducted to discover why rock formations fold in one place and fracture in another. Rocks can be classified as brittle or ductile on the basis of the way they behave when subjected to deforming forces. A brittle material breaks suddenly when its elastic limit is reached. A ductile material experiences smooth and continuous plastic deformation even after its elastic limit is reached. As we have seen, rocks behave differently depending on the pressures and temperatures of their environment. Marble, for example, is brittle at shallow depths but ductile deeper in the crust. Generally, most igneous rocks are stronger (less deformable) than most sedimentary rocks, and basement rocks (old igneous or metamorphic rocks) are more brittle than the ductile young sedimentary rocks that may overlie them.

ROCK DEFORMATION AND PLATE TECTONICS

Folds and faults are the evidence of deformation that geologists map in the field. They provide clues to the larger panorama of forces that result from plate tectonics.

    Compressional forces are associated with convergent plate boundaries, where plates (especially two continental plates) collide. In these areas, rocks are being pushed together. The result is uplift and folding of the rocks into mountain belts. Compressional forces can also fracture rocks, resulting in reverse faults.

    Tensional forces operate at divergent plate boundaries, where plates are being pulled apart. These forces result in normal faulting and down-faulted blocks, or rift valleys. Rift valleys may occur on the seafloor (at mid-ocean ridges) or where two continental plates are being pulled apart (such as the Red Sea rift between the African Plate and the Arabian Plate).

    Shearing forces occur at transform plate boundaries, where plates slide horizontally past each other. These forces result in strike-slip faults. A transform fault is a strike-slip fault that forms a plate boundary. A good example of a transform fault is the San Andreas fault, which forms the boundary where the North American Plate and the Pacific Plate slide past each other.

    Crustal forces can also be strong in the middle of plates and can cause faulting in rocks far from plate boundaries.

PLUNGING FOLDS

Plunging folds are sometimes hard to visualize. If you were to take a conical ice cream cone (one with a pointed end) and carefully cut it in half along its vertical axis, you would have two features that would resemble plunging folds. If they were both placed on their flat edge (the edge that was formed by the cut), they would represent plunging anticlines, because the uparched portion of the cone is in the form of an anticline. If you could carefully balance the cone halves on their rounded edges, these would represent plunging synclines. The limbs of a plunging anticline become smaller and begin to close in the direction of plunge, whereas the limbs of a plunging syncline open up and become larger in the direction of plunge.

♦   Layered rocks can fold in several basic ways in response to compressional forces, depending on the properties of the rocks and the details of the forces. Upfolds, or arches, are called anticlines; downfolds, or troughs, are called synclines. The two sides of a fold are its limbs. The axial plane is an imaginary surface that divides a fold as symmetrically as possible, with one limb on either side of the plane. The line made by the lengthwise interaction of the axial plane with the beds is the fold axis. If the fold axis is not horizontal, the fold is called a plunging fold. *(See the accompanying article on plunging folds.)*

Not every fold has a vertical axial plane with limbs dipping symmetrically away from the axis. An asymmetrical fold is one in which one limb dips more steeply than the other. An overturned fold results when one limb has been tilted beyond the vertical. Both limbs of an overturned fold dip in the same direction, but the order of the layers in the bottom limb is the reverse of their original sequence: older rocks are on top of younger rocks.

If you follow the axis of any fold in the field, eventually the fold dies out and appears to plunge into the ground as it disappears. Such folds are called plunging anticlines and plunging synclines.

A dome is a broad circular or oval upward bulge of rock layers resembling a short anticline. The flanking beds encircle a central point and dip radially away from it. A basin is a bowl-shaped depression of rock layers resembling a short syncline. The flanking beds dip radially toward a central point. Some domes are probably created by igneous rock intruding the crust and pushing the overlying sediments upward. Some basins are formed when a heated portion of the crust cools and contracts, causing the overlying sediments to subside; others result when the crust is stretched by tectonic forces.

♦   Observations in the field seldom provide geologists with complete information. Bedrock may be obscured by overlying soils, or erosion may have removed evidence of former structures. So geologists search for clues they can use to work out the relationship of one bed to another. For example, an eroded anticline might be recognized by a strip of older rocks forming a core bordered on both sides by younger rocks dipping away.

Folds typically occur in elongated groups. Such a fold belt suggests that the region was compressed by horizontal tectonic forces. A good example is the Valley and Ridge fold belt of the Appalachian Mountains.

♦   There are two kinds of fractures: a joint is a crack along which no appreciable movement has occurred; a fault is a fracture with relative movement of the rocks on both sides of it, parallel to the fracture.

Joints are found in almost every outcrop. Brittle rocks break more easily at weak spots when they are subjected to pressure. Regional tectonic forces that have long since vanished may leave their imprint in the form of a set of joints. Joints can also result from expansion and contraction of rocks when erosion has stripped away surface layers. Joints can form in lava as a result of the contraction of the lava as it cools. When a formation fractures and develops joints, this is usually the beginning of a series of changes that will alter the formation. For example, joints provide channels through which water and air can penetrate the formation and speed weathering.

♦   Faults can be caused by all three types of forces: compressive, tensional, and shearing. Faults are often found in mountain belts, where plates collide, and in rift valleys, where plates are pulled apart. Transform faults result where plates slide past each other. Crustal forces within plates can cause faulting in rocks far from plate boundaries.

Faults are defined by the direction of relative movement, or slip, at the fracture. The surface along which the formation fractures and slips is the fault plane. The orientation of the fault plane is described by strike and dip. A dip-slip fault involves relative movement of the formations up or down the dip of the fault plane. A strike-slip fault is one in which the movement is horizontal, parallel to the strike of the fault plane. A movement along the strike and simultaneously up or down the dip is an oblique-slip fault. Dip-slip faults are associated with compression or tension, strike-slip faults with shear. An oblique-slip fault suggests a combination of the two.

In a normal fault, the rocks above the fault plane move down in relation to the rocks below the fault plane, causing an extension of the section. A reverse fault is one in which the rocks above the fault plane move upward in relation to the rocks below, causing a shortening of the section. Reverse faulting results from compression. If, as we face a strike-slip fault, the block on the other side is displaced to the right, the fault is a right-lateral fault; if the block on the other side of the fault is displaced to the left, it is a left-lateral fault. These movements result from shearing forces. Finally, a thrust fault is a reverse fault at which the dip of the fault plane is small, so that the overlying block is pushed mainly horizontally. Thrust faults at which one block has been pushed horizontally over the other are called overthrusts and are often found in intensely deformed mountain belts.

   Tensional forces may split a plate apart, resulting in the development of a rift valley—a depression where one block looks as though it had dropped between two flanking blocks that have been pulled apart.

◆   Today's high mountain ranges can be traced to deformation that occurred over the past few tens of millions of years. Deformation that occurred hundreds of millions of years ago, however, no longer shows as prominent mountains. Erosion has left behind only the remnants of folds and faults in the old basement rocks of the continental interior. Geologists must decipher these remnants to reconstruct the geologic history of the region.

CHAPTER 10 OUTLINE

TYPES OF FAULTS

Describe each of the three types of faults depicted below. Address how they differ from one another, give the direction of the relative movement, or slip, at the fracture, and identify the type or types of tectonic forces associated with each.

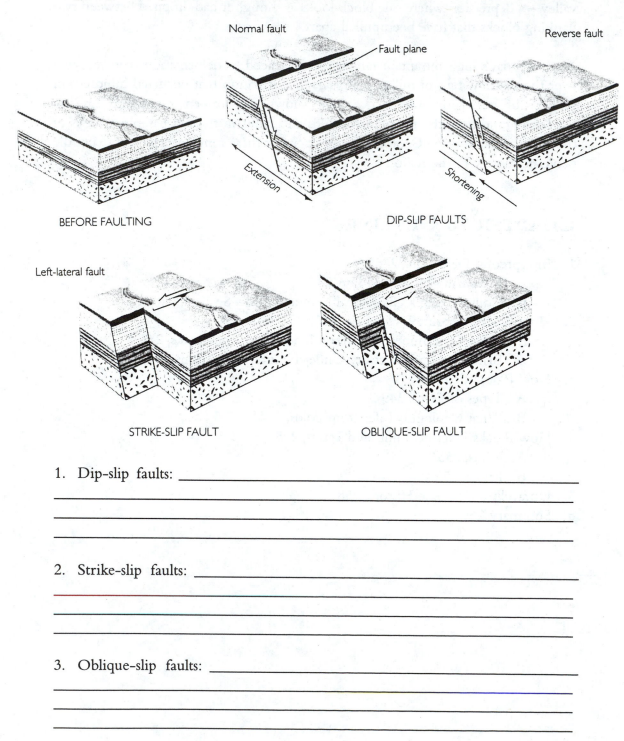

1. Dip–slip faults: _____

   _____

   _____

   _____

2. Strike–slip faults: _____

   _____

   _____

   _____

3. Oblique–slip faults: _____

   _____

   _____

   _____

PRACTICE MULTIPLE-CHOICE QUESTIONS

Circle the option that best answers the question.

1. Most deformation in Earth's crust is caused by:
   (a) Folding
   (b) Faulting
   (c) Seafloor spreading
   (d) Tectonic forces

2. The two measurements that define the orientation of an exposed rock layer are:
   (a) Strike and slip
   (b) Slip and dip
   (c) Strike and dip
   (d) Limb and fold axis

3. A fault plane that strikes to the northwest could be dipping either to the:
   (a) Northeast or southwest
   (b) Northeast or southeast
   (c) Northwest or southeast
   (d) Northwest or southwest

4. A rock that breaks suddenly when its elastic limit is reached is:
   (a) Elastic
   (b) Plastic
   (c) Brittle
   (d) Ductile

5. An elastic material:
   (a) Exists only under high pressure
   (b) Is one that breaks under stress
   (c) Is one that returns to its original dimensions when stress is released
   (d) Is one that does not deform under stress

6. Rocks that are ductile tend to be associated with:
   (a) Faults
   (b) Folds
   (c) Strike-slip motion
   (d) Joints

7. An overturned fold is one in which:
   (a) Both limbs dip in different directions
   (b) The axial plane is vertical
   (c) The axial plane is inclined
   (d) The strata in one limb are horizontal

8. In an anticline, the limbs dip:
    (a) Away from the axial trace and the youngest rocks are at the center
    (b) Away from the axial trace and the oldest rocks are at the center
    (c) Toward the axial trace and the oldest rocks are at the center
    (d) Toward the axial trace and the youngest rocks are at the center

9. A structure in which the beds dip away from a central point in all directions is:
    (a) A basin
    (b) A dome
    (c) An anticline
    (d) A syncline

10. If we find a rock in the field that has a crack along which only separation has occurred, we are looking at a:
    (a) Fault
    (b) Fold
    (c) Graben
    (d) Joint

11. A high-slip-angle (60∞) fault in which the rocks above the fault plane have moved down relative to the rocks below the fault plane is a:
    (a) Normal fault
    (b) Reverse fault
    (c) Thrust fault
    (d) Strike-slip fault

12. The formation of reverse faults is due to:
    (a) Compressional forces
    (b) Tensional forces
    (c) Soft, malleable rock
    (d) Brittle, nonelastic rock

13. The sense of motion along the San Andreas fault in California is:
    (a) Left-lateral strike slip
    (b) Right-lateral strike slip
    (c) Purely dip slip
    (d) Unknown due to a lack of information

14. If a continental area undergoes extension, such as has occurred in eastern Africa, the resulting feature is termed:
    (a) An overthrust
    (b) A transform fault
    (c) A rift valley
    (d) A plunging anticline

CD RESOURCES

SUPPLEMENTAL DIAGRAMS

10.30 Strike and dip of an inclined bed
10.31 Various fold orientations
10.31 Evolution of an overturned fold
10.32 A: Chevron folds. B: Kink folds. C: Chevron folds in sequence of alternating layers

SUPPLEMENTAL PHOTOS

10.34 Small-scale folds in interbedded shales and cherts
10.35 Tilted beds
10.36 Folded sedimentary rocks in Canadian Rockies
10.37 Anticline in Pennsylvanian rocks
10.38 Anticline; Boulder River, Montana
10.39 Syncline in limestone beds
10.40 Syncline in weathered shale
10.41 Syncline in Silurian rocks
10.42 Recumbent fold in Scapegoat Mountain, Montana
10.43 Recumbent fold
10.44 Dome at Sinclair, Wyoming
10.45 Columnar jointing in an andesitic flow
10.46 Cross section of joints
10.47 Columnar jointed basalt
10.48 Eroded joints in Arches National Park, Utah
10.49 Reverse fault, showing the fault drag on the right side
10.50 Reverse fault. Franciscan Formation
10.51 Drag fault in lake sediments
10.52 Right-lateral horizontal movement of San Andreas fault in 1906 San Francisco earthquake
10.53 Aerial view of Montagua transform fault, Guatemala
10.54 Montagua fault, Guatemala, from the ground

FIRST EDITION SLIDE SET

Slide 10.1 Drag fold along fault, Santa Catalina Mountains, Arizona
Slide 10.2 Faulted dike, Santa Catalina Mounts, Arizona
Slide 10.3 Wildrose graben, Panamint Valley, California
Slide 10.4 Fault scarp (1872) in Owens Valley, California
Slide 10.5 Aerial view of the Raplee anticline, Utah
Slide 10.6 Raplee anticline from the San Juan River, Utah
Slide 10.7 Overturned fold, Israel
Slide 10.8 Landers fault scarp, Mojave Desert, California
Slide 10.9 Joint-controlled landscape, Rainbow Plateau, Utah

SECOND EDITION SLIDE SET

Slide 10.1 Shatter cones in rock at Gosses Bluff meteor crater, Australia

Slide 10.2 Strike valley, Waterpocket Fold, Capitol Reef National Park, Southeastern Utah

Slide 10.3 Rattlesnake Mountain, west of Cody, Wyoming

Slide 10.4 The Lewis Thrust, Sawtooth Range, Montana

Slide 10.5 The Sawtooth Range in the Bob Marshall Wilderness, Montana

Slide 10.6 Thrust fault and repeated sedimentary layers in Ram's Horn Peak, Hoback Range, Wyoming

Slide 10.7 The Lander Dome Oil Field, Wind River Basin, Wyoming

Slide 10.8 Pirate Fault and Pusch Ridge, Santa Catalina Mountains, southeastern Arizona

Slide 10.9 Breccia in the Pirate Fault, Santa Catalina Mountains, southeastern Arizona

EXPANSION MODULES

The expansion module chapters on Canada and regions of the United States (Chapters 25–30) show numerous photographs of folding and faulting. Chapter 31, "The Himalayas and Tibetan Plateau," also contains several photographs of deformation.

ANIMATIONS

20.1   Plate Tectonics: Plate Divergence
20.2   Plate Tectonics: Continent-Continent Convergence
20.3   Plate Tectonics: Ocean-Continent Convergence
20.4   Plate Tectonics: Transform Boundary

ILLUSTRATED GLOSSARY

EXERCISES

Tectonic Forces in Rock Deformation
Types of Folds

CD LAB

1. The Raplee anticline was formed by:
    (a) Tensional forces during the Colorado orogeny
    (b) Compressional forces during the Colorado orogeny
    (c) Tensional forces during the Laramide orogeny
    (d) Compressional forces during the Laramide orogeny

    (Hint: See the first edition Slide Set.)

2. The French Thrust in Montana superposes:
    (a) Mississippian limestone over Cretaceous shale
    (b) Cretaceous shale over Mississippian limestone
    (c) Mississippian shale over Cretaceous limestone
    (d) Cretaceous limestone over Mississippian shale

(Hint: See the second edition Slide Set.)

3. The youngest Precambrian orogen of Laurentia is the:
    (a) Grenville orogen
    (b) Wopmay orogen
    (c) Thelon orogen
    (d) New Quebec orogen

(Hint: See Expansion Module Chapter 25.)

4. Intense folding, faulting, and metamorphism occur in the Himalayas. These processes are a result of the ongoing:
    (a) Collision of Tibet with Asia
    (b) Collision of India with Asia
    (c) Collision of Indochina with Tibet
    (d) Subduction of Asia under India

(Hint: See Expansion Module Chapter 31.)

5. Because the crust of both continents involved in the Himalayan orogeny is light and buoyant, the crust at the boundary:
    (a) Thins, as one plate is subducted beneath the other and melted
    (b) Thickens, because neither plate can be completely subducted
    (c) Remains the same thickness

(Hint: See Animation 20.2.)

Web Project: Are you interested in learning more about structural geology? If so, open your web browser and check out the <u>Chapter 10</u> link (reached from the WebNotes <u>By Chapter</u> link on the *Understanding Earth* home page). In case you've forgotten it, the home page is located at:

http://www.whfreeman.com/understandingearth

(Maybe you should make a bookmark for this page.)

There are several interesting links. For example, to look at lots of photographs of various types of deformation, click the <u>Geology 41</u> link, then the <u>Structure</u> link from that page. If you'd like to know more about how structural geologists work and the tools they use, try the <u>Structural Geology Tutorials</u>. Don't be daunted by the fact that the second Introductory Tutorial is on trigonometry. All of these introductory tutorials are presented very well and are quite easy to understand. (And if you want to explore a career in structural geology, you'll get some idea of the kinds of skills you'll need to have.)

FINAL REVIEW

After reading this chapter, you should:

- Know the forces involved in rock deformation and how they differ
- Know the parts of folds and the various types of folds
- Understand how joints and faults occur
- Know the differences between types of joints and faults
- Know how geologic evidence of rock deformation can be used to determine geologic history

ANSWER KEY

TYPES OF FAULTS

1. A dip-slip fault involves relative movement of the formations up or down the dip of the fault plane. In a normal fault, the rocks above the fault plane move down in relation to the rocks below the fault plane; in a reverse fault, the rocks above the fault plane move up in relation to the rocks below the fault plane. Normal faults result from tensional forces; reverse faults from compressional forces.
2. A strike-slip fault involves horizontal movement parallel to the strike of the fault plane. Strike-slip faults result from shearing forces.
3. An oblique-slip fault involves movement simultaneously along the strike and up or down the dip of the fault plane. Oblique-slip faults result from a combination of tensional or compressional forces and shearing forces.

PRACTICE MULTIPLE-CHOICE QUESTIONS

1. d, 2. c, 3. a, 4. c, 5. d, 6. b, 7. c, 8. b, 9. b, 10. d, 11. a, 12. a, 13. b, 14. c

Mass Wasting

This chapter focuses on the downhill movements of masses of soil, rock, or mud. Mass movements occur when the force of gravity exceeds the strength of the materials on a slope. Such movements can displace small amounts of soil down a gentle hillside or be huge landslides that dump tons of earth and rock on valley floors below steep mountains.

Mass wasting includes all the processes by which masses of rock and soil move downhill under the influence of gravity, eventually to be carried away by other transporting agents. Mass wasting is one result of weathering and rock fragmentation, and it is an important part of the general erosion of the land, especially in hilly and mountainous areas.

Human interference can have profound effects on mass movements. In the United States alone, for example, excavation for houses and buildings breaks up and transports up to 800 million tons of surface materials, far more than the 600 million tons moved annually by natural processes.

♦ Three primary factors influence mass movements: the nature of the slope materials; the amount of water in the materials; and the steepness and stability of slopes. Slope materials differ greatly in different kinds of terrain because they depend on the local geology. The metamorphic bedrock of one hillside may be badly fractured by foliation, while another slope nearby may be composed of massive granite.

♦ The angle of repose is the maximum angle at which a slope of loose (unconsolidated) material will lie without cascading down. A slope that is steeper than the angle of repose is unstable and will tend to collapse to the stable angle. The angle of repose varies with several factors, one of which is the size and shape of the particles. Larger, flatter, and more angular pieces of loose material remain stable on steeper slopes. The angle of repose also varies with the amount of moisture between particles: that of damp sand is higher than that of dry sand because the small amount of moisture between the grains binds them together through surface tension (the attractive force between molecules at a surface). But too much water keeps the particles apart and allows them to move freely over one another. Saturated sand, in which all the pore space is occupied by water, runs like a fluid.

♦   Consolidated (compacted and cemented) dry materials do not have the simple angles of repose characteristic of loose materials. The slopes of consolidated materials may be steeper and less regular, but they can become unstable when they are oversteepened or denuded of vegetation. Both cohesive and adhesives forces bind together the particles of consolidated dry materials. Cohesion is an attractive force between particles of a solid material that are close together. Adhesion is an attractive force between materials of different kinds.

Mass movements of consolidated materials usually result from the effects of moisture, often in combination with other factors. When the ground becomes saturated with water, the material is lubricated and the particles or larger aggregates can move past one another more easily. Soils become more susceptible to erosion and mass movement when they are stripped of vegetation. When the soil is no longer bound by root systems, the slope becomes more vulnerable to water invasion.

♦   Rock slopes range from gentle inclines of easily weathered rocks to vertical cliffs of hard rocks such as granite. The stability of these slopes depends on the weathering and fragmentation of the rock. Shales, for example, tend to weather and fragment into small pieces that form a thin layer of loose rubble covering the bedrock. The resulting slope angle of the bedrock is similar to the angle of repose of loose, coarse sand. Limestones and hard, cemented sandstones in arid environments resist erosion and break into large blocks, resulting in steep, bare bedrock slopes above and gentler slopes covered with broken rock below.

The structure of rock beds influences their stability, especially when the dip of the beds parallels the angle of the slope. Bedding planes may be zones of potential weakness because the adjacent beds differ in their mineral composition and texture or in their ability to absorb water. Such beds may become unstable, allowing masses of rock to slide along the weak bedding planes.

♦   A number of factors can trigger a mass movement. Sometimes a slide or debris flow is provoked by a heavy rainstorm. Many slides are set off by earthquake vibrations. Others may be precipitated by no single detectable event but just by gradual steepening that eventually results in the slope suddenly giving way.

♦   Geologists recognize many different kinds of mass movements, each with its own characteristics. Mass movements are classified by the nature of the material, the speed of the movement, and the nature of the movement (for example, sliding or flowing).

♦   Rock movements include rockfalls, rockslides, and rock avalanches. During a rockfall, individual blocks broken from the outcrop by chemical and physical weathering and erosion plummet suddenly in free-fall from a cliff or steep mountainside. Rockslides are the fast movements of large masses of bedrock sliding more or less as a unit, often along downward-sloping bedding or joint planes. Rock avalanches are composed of large masses of rocky materials that have broken up into smaller pieces by falling and sliding and then flow farther downhill at velocities of tens to hundreds of kilometers per hour. Rock avalanches are some of the most destructive mass movements.

♦   Most unconsolidated mass movements are slower than most rock movements. The slowest unconsolidated mass movement is creep, the downhill movement of soil or other debris at a rate of about 1 to 10 mm per year. Earthflows and debris flows are fluid mass movements that travel at up to a few kilometers per hour. Earthflows are fluid movements of relatively fine-grained materials, such as soils, weathered shales, and clay. Debris flows are fluid mass movements of rock fragments supported by a muddy matrix. They contain much material coarser than sand and tend to move more rapidly than earthflows. Mudflows are flowing masses of material mostly finer than sand, along with some rock debris, containing large amounts of water. Many mudflows move at several kilometers per hour. Fastest of all unconsolidated flows are debris avalanches, rapid downhill movements of soil and rock that usually occur in humid mountainous regions. Their speed is due to the combination of high water content and steep slopes. Water-saturated debris may move as fast as 70 km per hour.

A slump is a slow slide of unconsolidated material that travels as a unit. A debris slide involves the movement of rock material and soil largely as one or more units along planes of weakness, such as a waterlogged clay zone within or at the base of the debris.

Solifluction occurs only in cold regions when water in the surface layers of the soil alternately freezes and thaws. When the surface layers thaw, the soil there becomes waterlogged. The water cannot seep downward because the deeper layers of soil are still frozen, so it continues to accumulate and saturates the upper layers of soil so thoroughly that they ooze downhill, carrying broken rocks and other debris with them.

♦   Materials from mass movements erode easily and extensively because they have already been broken into finer grain sizes and larger surface areas. When these weathered materials reach lower slopes, they are transferred to small streams. Thus, few mass movements are preserved in the ancient rock record. It is unusual to find mass movements millions of years old.

♦   Although no one has yet witnessed a slide underwater, we know from slide materials found on the seafloor that submarine slides do occur. One such slide was caused by the weakening of an undersea slope at the Mid-Atlantic Ridge rift valley by faulting and hydrothermal metamorphism. Marine volcanic slides can result from the sudden eruption of a submarine volcano or of a land volcano bordering the ocean.

◆   To find the causes of modern slides, geologists correlate eyewitness reports with geological investigations of the source of the slide and the distribution and nature of the debris dropped in the valley below. Causes of prehistoric slides can be inferred from geological evidence alone where the debris is still present and can be analyzed for size, shape, and composition.

◆   There is hardly an area of geology that plate tectonics has not touched, and mass wasting is no exception. We can make the connection between mass movements and plate tectonics by considering topography, slopes, surface material composition, and rock-weakening processes. *(See the accompanying article on mass wasting and plate tectonics.)*

MASS WASTING AND PLATE TECTONICS

Mass wasting is likely to be found at the boundaries where two plates converge and mountains are uplifted. At the convergence of an oceanic plate with a continental plate and the associated subduction zone (such as the Andes Mountains along the western coast of South America), we find high, steep mountains slopes and frequent volcanic eruptions. As we have seen, steep slopes contribute to mass movements, and volcanic eruptions produce ash that can easily become mudflows. In addition, the fracturing and deformation of rocks that occur during mountain building and the many earthquakes along subduction zones contribute to frequent mass movements in these settings. A continent-continent plate collision, such as the Himalayas, meets all of the same conditions except for volcanism.

Mass movements also occur at divergent plate boundaries, particularly on the steep slopes formed in continental rift valleys. Submarine slides can occur at ocean rift valleys, such as the central rift valley of the Mid-Atlantic Ridge.

Transform plate boundaries are also frequently the sites of mass movements. The steep slopes that may be found along the fault and the prevalence of earthquakes contribute to mass movements in these settings. A good example is the San Andreas fault in California.

The conditions in all of these plate-tectonic settings contrast with those at areas far from present or former plate boundaries. In such settings, the topography is relatively low, the rocks are relatively undisturbed, the slopes are gentle, and earthquakes are rare.

CHAPTER 11 OUTLINE

CLASSIFICATION OF MASS MOVEMENTS

Complete the table below by supplying the correct mass movement in the numbered blanks.

MATERIAL	MOTION	SLOW		MODERATE	FAST
			VELOCITY		
Rock	Flow				1
	Slide or fall			2	3
Unconsolidated	Flow	4	5	6	7
		8	9		
	Slide or fall		10	11	

1. _____
2. _____
3. _____
4. _____
5. _____
6. _____
7. _____
8. _____
9. _____
10. _____
11. _____

PRACTICE MULTIPLE-CHOICE QUESTIONS

Circle the option that best answers the question.

1. The angle of repose for sedimentary particles:
    (a) Increases as the height of the pile of sediment increases
    (b) Decreases as the height of the pile of sediment increases
    (c) Increases as the angularity and coarseness of the grains increase
    (d) Decreases as the angularity and coarseness of the grains increase

2. The particles that would have the lowest stable slope angle would be:
    (a) Angular boulders
    (b) Fine sand
    (c) Coarse sand
    (d) Angular pebbles

3. Damp sand has a higher angle of repose than dry sand because of:
    (a) Adhesion
    (b) Cohesion
    (c) Surface tension
    (d) Cementation

4. The three primary factors that influence mass movements are:
    (a) The nature of the materials, the amount of water present, the slope steepness and stability
    (b) The nature of the materials, the amount of water present, the plate-tectonic setting
    (c) The amount of water present, the slope steepness and stability, the plate-tectonic setting
    (d) The nature of the materials, the slope steepness and stability, the plate-tectonic setting

5. The fast movement of large masses of bedrock sliding more or less as a unit is termed a:

    (a) Rockfall
    (b) Rockslide
    (c) Rock avalanche
    (d) Slump

6. Material that accumulates at the base of a steep hill or cliff is termed:
    (a) Outwash
    (b) Talus
    (c) Slump debris
    (d) Slide debris

7. A mass movement that is a mixture of unconsolidated material and has a high velocity is a:
    (a) Rockslide
    (b) Landslide
    (c) Slump
    (d) Debris avalanche

8. The type of mass movement in which the material moves downward and along a curved plane as a unit is termed a:
    (a) Slump
    (b) Rockslide
    (c) Debris flow
    (d) Debris avalanche

9. The large amount of energy that moves an earthflow or a debris avalanche comes from:
    (a) Solar heat
    (b) The angle of the slope over which the material moves
    (c) Gravity
    (d) Water, which is the lubricant

10. The type of mass movement that is confined to cold regions is:
    (a) Solifluction
    (b) Rock avalanche
    (c) Slump
    (d) Earthflow

11. Evidence for the occurrence of slumping is seen in:
    (a) Large piles of rubble at the foot of a hill
    (b) The downhill tilting of telephone poles and fences
    (c) The presence of water flowing from the stop of the slump
    (d) Blocks of solid rock being moved upslope

12. The text discusses the Gros Ventre slide that occurred in Wyoming in 1925. This slide occurred because:
    (a) The snow pack had built up and added too much weight for the rocks to hold back
    (b) A river had undermined the base of the cliff
    (c) A river had undermined the base of the cliff, and heavy rains fell, reducing the friction between a sandstone layer and a shale layer
    (d) An earthquake in the nearby Grand Tetons triggered the slide

CD RESOURCES

SUPPLEMENTAL PHOTOS

11.19   Rockfall on Interstate 70, Colorado
11.20   Talus slope
11.21   Avalanche chutes and talus cones, Tioga Pass, California
11.22   Rockslide in motion
11.23   Rockslide on Interstate 80, Echo Canyon, Utah
11.24   Rockslide on a highway
11.25   Scarp area at head of Kirkwood earthflow, Montana
11.26   Earthflow
11.27   Mudflow caused by the eruption of Mount Pinatubo
11.28   Slumgullion mudflow, Hinsdale County, Colorado
11.29   Fazenda Creek Valley mudflow at Nilo Pecanha power plant, Brazil
11.30   Lahar/mudflow, North Fork Toutle River, Mount St. Helens
11.31   Lahar/mudflow surface, Upper Muddy River fan, Mount St. Helens
11.32   Upper end of a debris avalanche
11.33   Remnant of a debris avalanche at Spirit Lake
11.34   Slump near Big Horn, Wyoming
11.35   Aerial view of slump in a hillslope
11.36   Impact of hillside creep on railroad, Yukon Territory, Canada
11.37   Creep in vertical layers of shale, Maryland
11.38   Solifluction lobes on the side of a kame in Greenland
11.39   Solifluction lobes, Greenland
11.40   Rupture surface of an ancient landslide
11.41   Landslide scarp developing in residential area
11.42   Landslide scarp after slide
11.43   Landslide in Los Angeles County, California
11.44   Landslide in Los Angeles County, California
11.45   Madison slide and Quake Lake, Montana
11.46   Gros Ventre, Wyoming, landslide
11.47   Gros Ventre, Wyoming, landslide in 1963

FIRST EDITION SLIDE SET

Slide 11.1 Boulders at the base of Vermilion Cliffs, Marble Canyon, Arizona
Slide 11.2 Avalanche, Banff National Park, Alberta, Canada
Slide 11.3 Mudflow, Toutle River, Washington

SECOND EDITION SLIDE SET

Slide 11.1 The Leaning Tower of Pisa and Duomo, Pisa, Italy

OTHER CHAPTERS

The supplemental photos for Chapters 5 (Volcanism) and 18 (Earthquakes) contain examples of mass movements associated with volcanoes and earthquakes. Expansion Module Chapter 25 (Geology of Canada) has photos of solifluction.

ILLUSTRATED GLOSSARY

EXERCISES

Rock Mass Movements
Unconsolidated Mass Movements

CD LAB

1. A rare photo of a rockslide in motion was taken at:
    (a) Mount Pinatubo, the Philippines
    (b) Mount St. Helens, Washington
    (c) Kilauea, Hawaii
    (d) Tioga Pass, California

    (Hint: See the Chapter 11 supplemental photos.)

2. Vermilion Cliffs in Marble Canyon, Arizona, are made up of alternating layers of erosion-resistant sandstone/conglomerate and weak shale/siltstones. Boulders that fall to the bottom of the cliffs are:
    (a) Sandstone and conglomerate
    (b) Shale and siltstones

    (Hint: See the first edition Slide Set.)

3. Slide 11.2 in the first edition Slide Set is a dramatic photo of an ice and snow avalanche in Banff National Park, Alberta, Canada. What is the substance that mixes with the snow and ice and increases its mobility?
    (a) Water
    (b) Air
    (c) Rocks
    (d) Mud

    (Hint: See the first edition Slide Set.)

FINAL REVIEW

After reading this chapter, you should:

- Know the three primary factors that affect mass movements
- Understand how mass movements are classified and know the types of mass movements
- Know how catastrophic mass movements are studied and what is gained from such studies
- Know what effects human activities can have on mass movements

ANSWER KEY

CLASSIFICATION OF MASS MOVEMENTS

1. Rock avalanche, 2. Rockslide, 3. Rockfall, 4. Creep, 5. Earthflow, 6. Mudflow, 7. Debris avalanche, 8. Solifluction, 9. Debris flow, 10. Slump, 11. Debris slide

PRACTICE MULTIPLE-CHOICE QUESTIONS

1. c, 2. b, 3. c, 4. a, 5. b, 6. b, 7. d, 8. a, 9. c, 10. a, 11. a, 12. c

The Hydrologic Cycle and Groundwater

Hydrology is the study of the movement and characteristics of water in and on the Earth. This chapter examines flows and reservoirs, the hydrologic cycle, hydrology and climate, precipitation and runoff, groundwater, aquifers, erosion by groundwater, water quality and resources, and hydrothermal water.

♦ Earth's water is stored in five major reservoirs: the oceans (by far the largest reservoir), glaciers and polar ice, groundwater (underground water), lakes and rivers, the atmosphere, and the biosphere. Figure 12.1 shows the distribution of water among these five reservoirs. Reservoirs gain water through rain and river inflow and lose water through evaporation and river outflow. If the inflow and outflow are equal, the size of the reservoir stays the same, even though water is constantly entering and leaving.

♦ The hydrologic cycle is the cyclical movement of water from the ocean to the atmosphere by evaporation, through rain to the surface, through runoff and groundwater to streams, and back to the ocean. Figure 12.2 shows the complete hydrologic cycle. Earth's external heat engine, powered by the Sun, drives the hydrologic cycle, evaporating water from the oceans and transporting it in the atmosphere as water vapor, which condenses to form clouds and eventually falls as rain or snow over the oceans and continents. Some of the water that falls on land infiltrates rock or soil through joints or small pore spaces between particles. Some groundwater evaporates through the soil surface, and some is returned to the atmosphere by transpiration, the release of water vapor from plants. Other groundwater is returned to the surface by springs that empty into rivers and lakes.

The rainwater that does not infiltrate the ground runs off the surface, gradually collecting into streams and rivers. The total of all rainwater that flows over the surface is called runoff.

Snowfall may be converted to ice in glaciers, which return the water to the oceans by melting and runoff and to the atmosphere by sublimation, which is the transformation from a solid (ice) directly to a gas (water vapor). Most of the water that evaporates from the oceans returns to them as precipitation. The remainder falls over the land and either evaporates or returns to the ocean as runoff.

◆   Ultimately, the global hydrologic cycle controls water supplies. We cannot use the fraction of precipitation that evaporates. We can use the fraction that becomes groundwater by digging wells. Runoff is the most readily available fraction. The most desirable water supplies are those that are rapidly and continually replenished by runoff and infiltration.

Human activities interfere with the natural hydrologic cycle. For example, evaporation is increased by the use of irrigation in dry areas, runoff patterns are altered when water is diverted from one region to another, and human contributions to global and local warming can lead to melting of glacial ice and changes in the balance of water in the other reservoirs.

◆   For practical purposes, local hydrology is more important than global hydrology. The strongest influence on local hydrology is the climate. Many differences in climate are related to the temperature of the air and its relative humidity—the amount of water vapor it contains. Warm air can hold much more water vapor than cold air, and most of the world's rain falls in warm humid regions near the equator. Landscape can alter precipitation patterns, as when a mountain range forms a "rain shadow," an area of low rainfall on the leeward (downwind) slopes. Polar climates tend to be very dry because the air and oceans are cold, which results in very low relative humidity. Between the tropical and polar extremes are the temperate regions, where rainfall and temperatures are moderate. Droughts—periods of months to years when precipitation is much lower than normal—can occur in any climate.

Climate also controls the relationship between precipitation and runoff. In areas of low precipitation, only a small part of the precipitation ends up as runoff. In humid regions, a much higher proportion of the precipitation runs off in rivers. Surface runoff collects in natural lakes, artificial reservoirs created by the damming of rivers, wetland areas, swamps, and marshlands. All these reservoirs absorb short-term inflows of major precipitation and release water during dry seasons or droughts. Thus, they are important in flood control because they smooth out seasonal or yearly variations in runoff and release steady flows downstream. Wetlands are fast disappearing as land development continues. In the United States, more than half the original wetlands are gone.

♦   Groundwater forms as rain infiltrates soil and other unconsolidated surface materials and even bedrock. Beds that store and transmit groundwater in sufficient quantity to supply wells are called aquifers.

As water moves into and through the ground, there are several factors that determine where and how fast it flows. One factor is porosity—the amount of pore space in rock, soil, or sediment. Porosity depends on the size and shape of the grains and how they are packed together. Large, loosely packed particles have higher porosity, and smaller particles that vary in shape have less porosity. Minerals that cement grains reduce porosity.

Another important factor is permeability—the ability of a solid to allow fluids to pass through it. Generally, but not always, permeability increases as porosity increases. Permeability depends on the pore sizes of the grains, how well the pores are connected, and how tortuous a path the water must travel to pass through the material.

♦   At shallower depths in the Earth, material is unsaturated (the pores contain some air and are not completely filled with water). This level is called the unsaturated zone. Below it is the saturated zone, the level in which the pores of the soil or rock are completely filled with water. The saturated and unsaturated zones can be in unconsolidated material or in bedrock. The groundwater table (or simply "water table") is the boundary between the two zones.

Water enters and leaves the saturated zone through recharge and discharge. Recharge is the infiltration of water into any subsurface formation. Influent streams, which recharge groundwater through the bottom of the stream where the stream channel lies above the water table, are most common in arid regions, where the water table is deep. Discharge is the exit of groundwater to the surface. When a stream channel intersects the water table, water discharges from the groundwater to the stream. Such an effluent stream is typical of humid areas.

♦   Groundwater may flow in unconfined or confined aquifers. An unconfined aquifer is one in which the water travels through permeable beds that extend to the surface, in both discharge and recharge areas. The level of the reservoir in an unconfined aquifer is the same as the height of the water table. A bed with low permeability that restricts the flow of groundwater is called an aquiclude. A con-fined aquifer is formed when aquicludes lie both over and under an aquifer.

The impermeable beds above a confined aquifer prevent rainwater from infiltrating downward into the aquifer. Instead, the confined aquifer is recharged by precipitation over the upland outcrop area that enters the ground and flows down the aquifer. Water in a confined aquifer—known as an artesian flow—is under pressure. If a well is drilled into a confined aquifer at a point where the elevation of the ground surface is lower than that of the water table in the recharge area, the water will flow out of the well spontaneously. Such wells are called artesian wells.

In some places, the water table may be more complicated. A perched water table can form where an aquiclude lies below the water table in a shallow aquifer and above the water table in a deep aquifer. Perched water tables range in size from small lenses to hundreds of square kilometers.

♦   When recharge and discharge are balanced, the groundwater reservoir and the water table remain constant. For recharge to balance discharge, there must be enough rainfall to equal the sum of the runoff from rivers and the outflow from springs and wells. But recharge and discharge are not always equal, because rainfall varies from season to season. Typically, the water table drops in drier seasons and rises during wet periods. A decrease in recharge, such as during a prolonged drought, is followed by a lowering of the water table.

An increase in discharge can also lower the water table. Shallow wells may end up in the unsaturated zone and dry up. When a well pumps water out of an aquifer faster than recharge can replenish it, the water level in the aquifer is lowered in a cone-shaped area around the well, called a cone of depression. If the cone of depression extends below the bottom of the well, that well goes dry.

♦   The balance between discharge and recharge is strongly affected by the speed at which water moves in the ground. Although all groundwaters flow through aquifers slowly, some flow more slowly than others. Darcy's law is an equation that can be used to determine the volume of water flowing between two points in a certain time when we know the vertical drop in the water table between the two points, the flow distance, and the permeability of the aquifer. The ratio between the elevation difference and the flow distance is known as the hydraulic gradient. Velocities calculated by Darcy's law have been confirmed experimentally. In most aquifers, groundwater moves at a rate of a few centimeters per day.

♦   Underground caves are produced when groundwater dissolves limestone or other soluble rocks. The amounts of limestone that have dissolved to make caves can be huge. As we saw in Chapter 6, the dissolution of limestone is enhanced by the atmospheric carbon dioxide contained in rainwater. As carbon dioxide–rich water moves down to the water table and the saturated zone, it dissolves carbonate minerals and creates openings. These openings enlarge as limestone dissolves along joints and fractures, forming a network of rooms and passages. We can explore caves that were once dissolved below the water table but are now in the unsaturated zone as a result of a drop in the water table. In these caves, water saturated with calcium carbonate may drip from the ceiling, accumulating into a long narrow spike of carbonate called a stalactite. As the remainder of the drop hits the floor, more carbon dioxide escapes and another small amount of calcium carbonate is precipitated on the cave floor below the stalactite. These deposits form a stalagmite. Eventually, a stalactite and stalagmite may grow together to form a column.

HYDROTHERMAL WATERS

Hydrothermal waters are hot waters found deep in Earth's crust. In some deep regions that are undergoing active metamorphism, such as along subduction zones, these waters play an important role in the chemical reactions of metamorphism, helping to dissolve some minerals and precipi-tating others.

Natural hot springs are found in many parts of the world. They occur where hydrothermal waters migrate upward so rapidly that they do not lose much of their heat, often emerging at the sur-face at boiling temper-atures. These waters are loaded with chemical substances dissolved from rocks at high temperatures. As hydrothermal waters come to the surface and cool quickly, they precipitate various minerals, such as opal (a form of silica) and calcite or aragonite (forms of calcium carbonate). Crusts of calcium carbonate that form at some hot springs build up to form the rock travertine.

Most hydrothermal waters of the continents come from surface waters that percolated downward to deeper regions of the crust. Hydro-thermal waters derived originally from rain or snow are called meteoric waters. Meteoric waters may be very old; for example, the water at Hot Springs, Arkansas, was derived from rain and snow that fell more than 4000 years ago.

The other source of hydro-thermal waters is water that escapes from a magma. In areas of igneous activity, sinking meteoric waters become heated by hot masses of rocks and then mix with water released from the nearby magma. The mixture of hydrothermal water returns to the surface as hot springs or geysers. The difference between hot springs and geysers is that hot springs flow steadily, whereas geysers erupt hot water and steam intermittently.

Geysers are probably connected to the surface by a system of very irregular and crooked fractures, re-cesses, and openings, in contrast to the more regular and direct plumbing of hot springs. The irregularity of the geyser plumbing, by isolating some water in recesses, helps to prevent the bottom waters from mixing with shallower waters and therefore cool-ing. As the bottom waters are heated by contact with hot rock, they reach the boiling point, which creates steam. The steam starts to rise and heat up the shallower waters. An eruption is triggered, followed by a return to quiet as the fractures slowly and irregularly refill with water.

Other hot springs come from meteoric waters that move downward into deep sedimentary rock formations, where they are heated by the normal increase in temperature with depth, and then return as hydrothermal waters to the surface. Many metallic ores and other mineral deposits in sedimentary rocks far from any igneous activity originated in this way.

Dissolution can thin the roof of a limestone cave so much that it collapses suddenly, producing a sinkhole. Sinkholes contribute to a distinctive form of topography known as karst. Karst topography occurs in regions with high rainfall, extensively jointed limestone formations, and appreciable hydraulic gradients.

♦ Almost all water supplies in North America are free of bacterial contamination, and the vast majority are chemically pure enough to drink safely. But contamination can occur in places where toxic wastes have polluted rivers and infiltrated aquifers from surface dumps. Lead contamination can be a problem in older homes where lead pipes are still common. Radioactive waste buried underground is another source of contamination. Some sources of toxic wastes include chemical industry waste lagoons and leaking chemical storage barrels; sanitary landfill operations and city garbage dumps; leaking buried gasoline tanks; road salt; agricultural pesticides, herbicides, and fertilizers; and septic tanks.

We can reverse the contamination of water supplies, but the process is costly and very slow. The faster an aquifer recharges, the easier it is to clean, because once the sources of contamination are removed, fresh water moves into the aquifer and the water recovers its quality in a short time. Contamination of slowly recharging reservoirs is a more serious problem. Even with cleaned-up recharge, some contaminated deep reservoirs hundreds of kilometers away from the recharge area may not respond for many decades.

♦ Other sources of water exist deep within the Earth. These waters move so slowly that they have plenty of time to dissolve even very insoluble minerals from the rocks they pass through. Thus, dissolved materials become more concentrated in these waters than in near-surface waters, rendering them undrinkable.

Hydrothermal water, which is associated with heat sources under the surface, contains large concentrations of dissolved minerals. When hydrothermal water reaches the surface, the dissolved minerals precipitate and form localized mineral deposits. *(See the accompanying article on hydrothermal waters.)*

Chapter 12 Outline

CONFINED AQUIFERS

Using the figure below, answer the questions that follow.

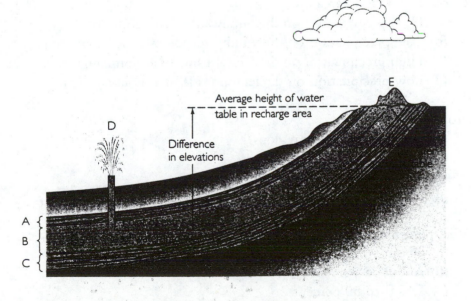

Points A and C are called (1) _____. Point B is a(n) (2) _____.
Point D is a flowing (3) _____. Point E is the upload (4) _____ area.

5. Where is the water entering the aquifer? _____ --
_____

6. What makes points A and C different from point B? _____

_____

7. What is causing the water to flow from point D? _____

_____

PRACTICE MULTIPLE-CHOICE QUESTIONS

Circle the option that best answers the question.

1. The smallest reservoir is:
    (a) Underground aquifers
    (b) The atmosphere
    (c) Lakes and streams
    (d) The biosphere

2. Water that does not infiltrate the ground in the hydrologic cycle is termed:
    (a) Runoff
    (b) Meteoric
    (c) Saturated
    (d) Magmatic

3. A rain shadow is an area of:
    (a) High precipitation on the windward side of a mountain range
    (b) Low precipitation on the windward side of a mountain range
    (c) High precipitation on the leeward side of a mountain range
    (d) Low precipitation on the leeward side of a mountain range

4. Droughts occur:
    (a) In arid or semiarid regions
    (b) In dry polar climates
    (c) In temperate climates
    (d) In any climate

5. What percentage of wetlands are left in the United States?
    (a) About 90 percent
    (b) About 75 percent
    (c) Less than 50 percent
    (d) Less than 20 percent

6. Porosity is the:
   (a) Number of pores in a rock
   (b) Volume of open space in a rock expressed as a percentage of the total rock volume
   (c) Ease with which water moves through the rock
   (d) Reaction of a rock to being saturated with water

7. The boundary separating the unsaturated zone from the saturated zone is the:
   (a) Cone of depression
   (b) Aquifer
   (c) Water table
   (d) Aquiclude

8. Good aquifers include all the following rocks except:
   (a) Granite
   (b) Sandstone
   (c) Fractured basalt
   (d) Gravel

9. Which of the following sets of geologic conditions must be met to have an artesian system?
   (a) An aquifer that is only underlain by an aquiclude
   (b) An aquifer that is only overlain by an aquiclude
   (c) An aquifer that is both overlain and underlain by aquicludes and an exposed recharge area for the aquifer
   (d) Highly fractured limestone or basalt units exposed on the surface

10. A perched water table would probably result in the presence of:
    (a) Artesian wells
    (b) Springs
    (c) New soil layers
    (d) Major rivers

11. Whenever too much water is pumped out of a well, what happens to the surface of the water table?
    (a) It remains horizontal
    (b) It parallels the topography of the ground surface
    (c) It dips down and forms a cone of depression
    (d) The water table actually rises up toward the surface

12. If a 30-foot-deep well is drilled in an area where the water table is situated 20 feet below the surface, the water will rise to what level in the well?
    (a) 5 feet
    (b) 10 feet
    (c) 20 feet
    (d) 30 feet

13. The lowering of the water table near a pumping well is called:
    (a) A cone of depression
    (b) Recharge
    (c) Capillary action
    (d) Artificial recharge

14. Darcy's law basically states that:
    (a) The size of sinkholes is related to the amount of subsurface water flow
    (b) Sinkhole spacing on the surface is related to topographic relief
    (c) Flow in an artesian system depends on the thickness of the aquicludes
    (d) The volume of water flowing in a given time is propor-tional to the ver-tical drop divided by the horizontal distance the water travels

15. The rate of flow of groundwater in the subsurface is on the order of:
    (a) Centimeters or maybe one or two meters per day
    (b) Centimeters or maybe one or two meters per year
    (c) Centimeters or maybe one or two meters per century
    (d) Several meters to about one kilometer per day

16. The formation of a cavern in limestone formations occurs:
    (a) By collapse of the rock underlying the cavern
    (b) Only when weak sulfuric acid is present
    (c) As a result of dissolution by slightly acidic circulating groundwater
    (d) In polar regions where limestone deposition is active

17. Dripstone deposits in caves that hang from the ceiling or overhead areas are:
    (a) Sinkholes
    (b) Stalactites
    (c) Stalagmites
    (d) Karst

18. Collapse associated with a series of underground caverns will produce which feature on the surface?
    (a) Sinkhole
    (b) Stalactite
    (c) Stalagmite
    (d) Perched water table

19. Hard water contains relatively large amounts of:
    (a) Magnesium and calcium ions
    (b) Iron and sulfur
    (c) Carbonic acid
    (d) Insoluble minerals

20. Surface waters that were originally derived from rain or snow are termed:
    (a) Vadose
    (b) Phreatic
    (c) Meteoric
    (d) Hydrothermal

21. The difference between a hot spring and a geyser is that:
    (a) A hot spring emits water at a lower temperature than a geyser
    (b) A hot spring emits water at a higher temperature than a geyser
    (c) A geyser erupts intermittently, whereas a hot spring flows steadily
    (d) A hot spring's plumbing system is more complicated than a geyser's

CD RESOURCES

SUPPLEMENTAL DIAGRAMS

12.25 The three states of matter as exemplified by water
12.26 Hydrologic cycle for the United States
12.27 Water consumption and supply in the United States

SUPPLEMENTAL PHOTOS

12.28 Marsh along shore of Lake Titicaca, in high plateau of Peru
12.29 Marshland along shore of Jackfish Lake, near Terrace Bay, Ontario
12.30 Scanning electron micrograph of a well-sorted, highly porous sandstone
12.31 Photomicrograph of a loosely cemented sandstone
12.32 Limestone cavern in southern Indiana
12.33 House lost in sinkhole, Polk County, Florida
12.34 Large sinkhole in central Alabama
12.35 Karst topography, Puerto Rico
12.36 Travertine deposits, Mammoth Hot Springs, Yellowstone

FIRST EDITION SLIDE SET

Slide 12.1 Thunder Springs, Grand Canyon, Arizona
Slide 12.2 Limestone cave, Bisbee, Arizona
Slide 12.3 Strokkur geyser, Geysir thermal area, Haukadalur, Iceland
Slide 12.4 San Pedro River, southeastern Arizona
Slide 12.5 Earth fissure, Picacho Basin, Arizona
Slide 12.6 Hoover Dam on the Colorado River, Arizona–Nevada state line

ILLUSTRATED GLOSSARY

EXERCISES

The Hydrologic Cycle

TOOLS

Calculator: Darcy's Law

CD LAB

1. Areas in which of the following states withdraw more than 100 percent of their renewable water supply?
    - (a) Nevada, Arizona, Texas
    - (b) California, Arizona, New Mexico
    - (c) Oregon, Colorado, New Mexico
    - (d) Oregon, Arizona, California

    (Hint: See the Chapter 12 supplemental diagrams.)

2. Which has a higher porosity?
    - (a) A well-sorted sandstone
    - (b) A loosely cemented sandstone

    (Hint: See Chapter 12 supplemental photos.)

3. Travertine deposits are composed of:
    - (a) NaCl
    - (b) H_2SO_4
    - (c) HCO_3
    - (d) $CaCO_3$

    (Hint: See Chapter 12 supplemental photos.)

4. Speleothems are:
    - (a) Rivers that run through caves and caverns
    - (b) Drops of moisture on the walls of caves and caverns
    - (c) Mineral deposits in caves, such as stalagmites and stalactites
    - (d) Small pools of water that form in underground caverns

    (Hint: See the first edition Slide Set.)

5. Geysers occur where:
    - (a) Water flows through rocks of high permeability and high thermal gradient
    - (b) Water flows through rocks of high permeability and low thermal gradient
    - (c) Water flows through rocks of low permeability and high thermal gradient
    - (d) Water flows through rocks of low permeability and low thermal gradient

    (Hint: See the first edition Slide Set.)

Web Project: Take a field trip and explore <u>The Virtual Cave</u>. (on the Chapter 12 page). Check out such fasci-nating cave features as "soda straws," "popcorn," "bot-tlebrushes," and "shower heads." Most people know about stalagmites and stalactites, but geologists have also discovered some unusual deposits called stegamites and splattermites. All these features and more await you in The Virtual Cave, accessed from the *Understanding Earth* home page:

http://www.whfreeman.com/understandingearth

FINAL REVIEW

After reading this chapter, you should:

- Thoroughly understand the workings of the hydrologic cycle
- Understand the relationship between precipitation and runoff
- Understand the basic dynamics of groundwater, including its importance, how it relates to porosity, the groundwater table, the workings of artesian flows, and recharge and discharge
- Know Darcy's law and understand its application to major aquifers
- Know the features of karst topography
- Know some of the issues involves in questions of water quality
- Know the properties of water deep in the crust

ANSWER KEY

CONFINED AQUIFERS

1. Aquicludes, 2. Confined aquifer, 3. Artesian well, 4. Recharge
5. At point E, through the aquifer in the upland recharge area
6. Permeability; the aquiclude has low permeability and the confined aquifer has high permeability
7. The artesian well flows in response to the difference in natural pressure (before the well was drilled) between the height of the water table in the recharge area and the bottom of the well

PRACTICE MULTIPLE-CHOICE QUESTIONS

1. d, 2. a, 3. d, 4. d, 5. c, 6. d, 7. c, 8. a, 9. c, 10. b, 11. c, 12. b, 13. a 14. d, 15. a, 16. c, 17. b, 18. a, 19. a, 20. c, 21. c

13

Rivers: Transport to the Oceans

Streams cover most of Earth's surface and play the dominant role in shaping the continental landscape. They erode mountains, carry the products of weathering to the oceans, and deposit billions of tons of sediment. Rivers carry most of the precipitation that falls on land back to the sea, completing the hydrologic cycle.

This chapter focuses on how water flows in currents; how currents carry and deposit sediment; how streams break up and erode solid rock; how streams carve valleys and assume a variety of forms as they channel water downstream; how streams change over time; and how and why streams flood. Any flowing body of water, large or small, is called a *stream;* the major branches of a large stream system are referred to as *rivers.*

♦ There are two basic types of fluid flow: laminar and turbulent. In laminar flow, straight or gently curved streamlines run parallel to one another without mixing or crossing. In turbulent flow, streamlines mix, cross, and form swirls and eddies. Fast-moving river waters typically show turbulent flow. Whether a flow is laminar or turbulent depends on its velocity, its geometry (primarily its depth), and its viscosity, or resistance to flow. The higher the viscosity, the greater the tendency for laminar flow. Because water has low viscosity in the range of temperatures found at Earth's surface, most streams in nature tend to be turbulent.

♦ Streams have different abilities to erode and carry sediment. Laminar flows can lift and carry only the smallest, lightest, clay-sized particles. Turbulent flows can move particles from clay size up to pebbles and cobbles. Turbulence lifts particles from the streambed into the flow and carries them downstream, rolling and sliding larger particles along the bottom. A stream's suspended load includes all the material temporarily or permanently suspended in the flow. Its bed load is the material it carries along the stream bottom by sliding and rolling.

The faster the current, the larger the particles carried as suspended and bed load. The ability of a flow to carry material of a given size is its competence. The total sediment load carried by a flow is its capacity.

♦   Turbulence tends to lift particles into the flow, and gravity tends to make them settle out of the current and become part of the bed. The settling velocity is the speed at which suspended particles of various weights settle to the bottom. Small particles settle slowly and tend to stay in suspension. Larger particles stay suspended in the current only a short time before they settle.

Sand grains typically move by saltation—an intermittent jumping motion along the streambed. The grains are sucked up into the flow by turbulent eddies, move with the current for a short distance, and then fall back to the bottom.

♦   When sand grains on a streambed are transported by saltation, they tend to form cross-bedded dunes and ripples. Dunes are elongated ridges of sand that can range up to many meters high. Ripples are very small dunes whose long dimension is formed at right angles to the current. As sand grains move by saltation, they are eroded from the upstream side of ripples and dunes and deposited on the downstream side, resulting in the slow downstream migration of these bedforms.

♦   Running water erodes solid rock by abrasion, by chemical and physical weathering, and by undercutting. The sand and pebbles carried by a river can wear away even the hardest rock by abrasion, and pebbles and cobbles rotating inside swirling eddies can grind deep potholes in the bedrock of the river bottom. Chemical weathering alters a rock's minerals and weakens it along joints and cracks. Physical weathering occurs as the impacts of boulders, pebbles, and sand split the rock along cracks.

♦   Streams create valleys as they erode the Earth's surface. A stream valley encompasses the entire area between the tops of the slopes on both sides of the river. The cross-sectional profile of river valleys may be V-shaped or may show a broad, low profile. At the bottom of the valley is the channel, the trough through which the water runs. In broader valleys, a floodplain—a flat area about level with the top of the channel—lies on either side of the channel.

In tectonically active, newly uplifted mountains, stream valleys are narrow and steep-walled, and the channel may occupy most or all of the valley bottom. In lowlands, where tectonic uplift has long since ceased, stream erosion of valley walls is helped by chemical weathering and mass wasting. Over time, these processes produce gentle slopes and floodplains many kilometers wide.

◆   Stream channels along the bottom of a valley may run straight in some stretches and show a meandering or braided channel pattern in others. Meanders—curves and bends in a channel—are typical of streams flowing on low slopes in plains or lowlands, where channels cut through unconsolidated sediments or easily eroded bedrock. Meanders are less pronounced but still common where the channel flows on higher slopes and harder bedrock. In such a terrain, meandering stretches may alternate with long, straight stretches.

Meanders on a floodplain migrate over time, eroding the outside banks of bends, where the current is strongest, and depositing point bars along the inside banks, where the current is slower. As meanders migrate, sometimes unevenly, the bends may grow closer and closer together, until finally the river bypasses the next loop. The river takes a new shorter course, and in its abandoned path it leaves behind a crescent-shaped, water-filled loop called an oxbow lake.

Braided streams have many channels that split apart and then rejoin, in a pattern resembling braids of hair. Braids are found in many settings, from broad valleys in lowlands to stream deposits in wide, downfaulted valleys adjacent to mountain ranges. They tend to form wherever rivers have large variations in volume of flow combined with a high sediment load and easily erodible banks.

◆   Channel migration over the floor of a valley creates a stream floodplain. Point bars and sediment deposited by floodwaters build up the surface of the floodplain. Erosional floodplains, covered with a thin layer of sediment, can form when a stream erodes bedrock or unconsolidated sediment as it migrates.

As floodwaters spread out over the floodplain, the velocity of the water slows and the current loses its ability to carry sediment. Because the speed of the flooding waters drops most quickly along the immediate borders of the channel, the current deposits coarse sediment, typically sand and gravel, along a narrow strip at the edge of the channel. Successive floods build up natural levees, ridges of coarse material that confine the stream within its banks between floods.

During floods, finer sediments are carried beyond the channel banks, often over the entire floodplain, and are deposited there. Fine-grained floodplain deposits form fertile soil, and many ancient and modern cities are located on floodplains.

◆   Streams are dynamic systems, moving from low to high waters and floods in a few years and reshaping their valleys over longer periods. Streams also change their flows and channel dimensions as they move downstream, from narrow valleys in their upland headwaters to broader floodplains in their middle and lower courses. Most of these longer term changes are adjustments in the normal volume and velocity of the flow and in the depth and width of the channel.

The size of a stream's flow is measured by its discharge—the volume of water that passes a given point in a given time as it flows through a channel of a certain width and depth. We find the discharge by multiplying the cross-sectional area by the velocity of the flow. Discharge typically increases downstream as more water flows in tributaries (streams that discharge water into larger streams). Increased discharge means that width, depth, or velocity must also increase.

♦   A flood is increased discharge that results from a short-term imbalance between input and output. As the discharge increases, the flow velocity in the channel increases and the water gradually fills the channel and then floods over the banks. Floods occur regularly, although at different intervals. Some are large, with very high water levels lasting for days; others are minor. Generally, small floods are more frequent, large floods less frequent.

The average time interval between the occurrence of two geological events of a given magnitude is called a recurrence interval. The recurrence interval of floods of different heights varies from stream to stream, depending on the climate of the region, the width of the floodplain, and the size of the channel.

♦   Over the long term, a stream is in dynamic equilibrium between erosion of the streambed and sedimentation in the channel and floodplain. This equilibrium is controlled by the topography (including slope), climate, stream flow (including both discharge and velocity), and resistance of the rock to weathering and erosion. *(See the accompanying article on longitudinal profile and grade.)*

♦   Every topographic rise between two streams, large or small, forms a divide— a ridge of high ground along which all rainfall is shed as runoff down one side or the other. All the divides that separate a stream and its tributaries from their neighbors define its drainage basin, an area of land surrounded by divides that funnels all its water into the network of streams draining the area.

The pattern of connections of the tributaries of a stream system is called a drainage network. Most rivers follow an irregular branching pattern called dendritic drainage. Dendritic drainage is typical of terrains where the bedrock is uniform, such as horizontally bedded sedimentary rocks or massive igneous or metamorphic rocks. A rectangular drainage pattern is dominated by right-angled bends, usually following fractures or joints in bedrock. A special rectangular pattern, trellis drainage, resembles the right-angled geometry of wooden trellises and forms when tributaries lie in the parallel valleys of steeply dipping beds in folded belts. Radial drainage occurs when a stream system runs in a radial pattern away from some central high point, such as a volcano.

Some drainage patterns are related to the geologic history of a region. An antecedent stream is one that was established before local uplift began and cut its channel at the same rate the land was rising. The stream channel is older than the geologic structure it cuts through. A superposed stream is one that was established on a new surface and that maintained its course despite different rocks and structures encountered as it cut downward into the underlying rocks.

LONGITUDINAL PROFILE AND GRADE

The slope of a river from headwaters to mouth can be described by plotting the elevation of its streambed against the distance from its headwaters. The resulting smooth, concave-upward curve is its longitudinal profile. All streams show this same general profile: steep near the stream's head and almost level near its mouth. The longitudinal profile is controlled at its lower end by a stream's base level, the elevation at which it enters a lake or the ocean and disappears as a stream.

Changes in natural base level affect the longitudinal profile. If the regional base level rises, the profile will show the effects of sedimentation as the river builds up channel and floodplain deposits to reach the new base-level elevation. Damming a river artificially can also create a new local base level, with similar effects on the longitudinal profile.

Over a period of years, a stream's profile becomes stable as the stream gradually reaches a balance between erosion and sedimentation. At equilibrium, the stream is a graded stream—one in which the slope, velocity, and discharge combine to transport its sediment load, with neither sedimentation nor erosion. If the conditions that produce a particular graded stream profile change, the stream's profile will change to reach a new equilibrium. Over geologic time, where regional base level is constant, the longitudinal profile reflects the balance between tectonic uplift and erosion on the one hand and transport and deposition on the other. If uplift is dominant, the profile is steep and expresses the dominance of erosion and transport. As uplift slows, the headwater region is eroded and the profile is lowered.

One place that a river must adjust suddenly to changed conditions is where it leaves a narrow mountain valley and enters a relatively broad, flat valley. In such locations, typically at steep fault scarps, streams drop large amounts of sediment in cone- or fan-shaped accumulations called alluvial fans. Coarse materials dominate on the steep upper slopes of the fan, finer sediments lower down.

When tectonic uplift changes the equilibrium of a stream valley, a flat steplike surface called a terrace may form above the floodplain. Terraces mark former floodplains that existed at the higher level before regional uplift or an increase in discharge caused the stream to erode into the former floodplain.

◆ Sooner or later, all rivers end as they flow into a lake or an ocean. As its current gradually dies out, a river loses its ability to transport sediment. The coarsest material, normally sand, is dropped first. Finer-grained sands are dropped farther out, followed by silt and then clay. As the floor of the lake or sea slopes to deeper water away from the shore, the different sizes of dropped materials build up a depositional platform called a delta. As the delta builds forward, the mouth of the river advances into the sea, leaving new land in its wake.

Strong waves, shoreline currents, and tides affect the growth and shape of deltas. Shoreline currents and waves can move the sediment along the shore, forming a long beach shoreline with only a slight seaward bulge at the mouth of the river. Where waves and tides are strong enough, deltas cannot form at all. Tectonics also exerts some control over where deltas form. Deltas require uplift in the drainage basin to provide abundant sediment and crustal subsidence in the delta region to accommodate huge deposits of sediment. Some large deltas are found on passive margins originally formed from a rifted continental margin. There are few large deltas associated with active subduction zones. The continental convergence that elevated the Himalayas also formed the great deltas of the Indus and Ganges.

CHAPTER 13 OUTLINE

MEANDERS

On the diagram of a meander below, draw a line running through the meander to indicate where the current is the strongest. Place E's at the points where erosion occurs and D's where deposition occurs.

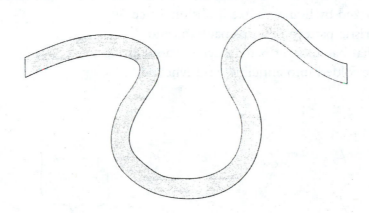

DRAINAGE PATTERNS

Match the four drainage patterns listed below with the lettered option that fits it—an option may fit more than one pattern.

Dendritic: _____

Rectangular: _____

Trellis: _____

Radial: _____

A. Typical of terrains where bedrock is uniform
B. Central uplift is a characteristic feature
C. Develops in valley and ridge terrain
D. Typical of rivers
E. Trunk streams are a feature
F. Rapid weathering along fractures in bedrock controls stream course
G. Characterized by branches like limbs on a tree
H. Characteristic pattern of a dormant volcano
I. Pattern that has another pattern type as a subgroup
J. Rocks are folded into anticlines and synclines

K.   L.   M.   N.

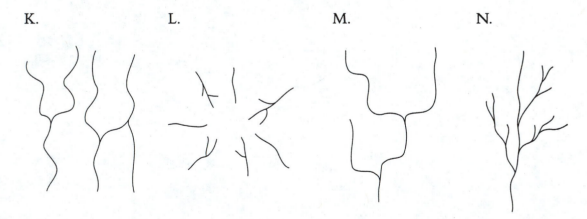

PRACTICE MULTIPLE-CHOICE QUESTIONS

Circle the option that best answers the question.

1. Most stream flows in nature tend to be:
   (a) Turbulent
   (b) Laminar
   (c) Turbulent only at the headwaters
   (d) Turbulent only near the mouth

2. The largest particles being moved by a stream will be part of the:
   (a) Dissolved load
   (b) Suspended load
   (c) Bed load
   (d) Overbank load

3. Stream competence is measured by:
   (a) The largest particle size that the stream can transport in its bed load
   (b) The amount of material in the dissolved load
   (c) Its maximum width along the floodplain
   (d) The total amounts of suspended load and bed load

4. Particles tend to settle out of the suspended load in the following order:
   (a) Clay, sand, pebbles
   (b) Pebbles, sand, silt
   (c) Sand, pebbles, cobbles
   (d) Clay, pebbles, sand

5. Sand grains in a stream typically move by:
   (a) Rolling and sliding along the bed
   (b) Suspension
   (c) Saltation
   (d) Turbulence

6. Which of the following statements about bedforms is true?
   (a) Ripples are larger than dunes
   (b) Dunes appear first, then ripples, then the bed becomes flat
   (c) Ripples and dunes tend to become smaller as current velocity increases
   (d) Ripples and dunes migrate downstream as current velocity increases

7. A river deepens its valley by:
   (a) Headward erosion
   (b) Lateral erosion
   (c) Downcutting
   (d) Deposition that produces point bars

8. A stream widens its channel by:
    (a) Downcutting
    (b) Depositing material to form point bars
    (c) Lateral erosion
    (d) Changing its longitudinal profile

9. Active erosion in a meander takes place:
    (a) In the center of the stream
    (b) Along the outer bank of a bend
    (c) Along the inside bank of a bend
    (d) Near a stream's headwaters

10. The eventual result of an ever-progressing meander is:
    (a) An alluvial fan
    (b) A point bar
    (c) A much narrower floodplain
    (d) An oxbow lake

11. Natural levees are built up on floodplains along the banks of rivers:
    (a) As floodwaters recede from the plain
    (b) As floodwaters rise and overflow the river banks
    (c) During droughts, when the water level is low
    (d) Continuously as normal discharge conditions prevail

12. A stream's discharge is:
    (a) Measured at the headwaters or source
    (b) The volume of water flowing past a given point in a specified time
    (c) Dependent on the amount of dissolved load
    (d) Measured by the amount of large particles in the bed load

13. The steepest portion of a stream's longitudinal profile is located:
    (a) Near the headwaters
    (b) In the central part of the stream
    (c) Just before the mouth
    (d) On the delta

14. Which of the following is an example of local base level?
    (a) A lake
    (b) The ocean
    (c) A point bar
    (d) A floodplain

15. An irregular branching drainage pattern is called:
    - (a) Radial
    - (b) Rectangular
    - (c) Trellis
    - (d) Dendritic

16. A stream that existed before the present topography was created and maintains its course despite changes in the underlying rocks is called:
    - (a) A dendritic stream
    - (b) A superposed stream
    - (c) An antecedent stream
    - (d) A channelized stream

17. Which of the following drainage patterns will develop on a volcano?
    - (a) Trellis
    - (b) Rectangular
    - (c) Dendritic
    - (d) Radial

18. Which of the following statements about deltas is true?
    - (a) Deltas build up wherever a river enters a lake or ocean
    - (b) Deltas require uplift in the drainage basin and subsidence in the delta area
    - (c) Many large deltas are associated with plate subduction zones
    - (d) Deltas remain roughly the same size over time

CD RESOURCES

SUPPLEMENTAL DIAGRAMS

13.29 Stream competence and capacity
13.30 Formation of valley deposits by a meandering channel
13.31 Formation of alluvial fan as stream adjusts its profile
13.32 Large rivers of the United States

SUPPLEMENTAL PHOTOS

13.33 Resistant rock structures along a stream channel
13.34 Potholes in granite, James River channel, Virginia
13.35 Palouse Falls exposes Columbia River basalts, Washington
13.36 Narrow, sharp-walled stream valley in high mountains
13.37 Stream valley with gentle slopes and broad floodplains in lowlands
13.38 Meandering stream with well-defined floodplain

FIRST EDITION SLIDE SET

ANIMATIONS

ILLUSTRATED GLOSSARY

EXERCISES

Marine Delta

TOOLS

Calculator: Stream Discharge

CD LAB

1. Point bars are usually:
   - (a) Horizontally bedded
   - (b) Cross-bedded
   - (c) Rippled

   (Hint: See Chapter 13 supplemental diagrams.)

2. The average flow of the Yukon River, Alaska, is:
   - (a) 160,000 ft^3/s
   - (b) 200,000 ft^3/s
   - (c) 240,000 ft^3/s
   - (d) 280,000 ft^3/s

   (Hint: See Chapter 13 supplemental diagrams.)

3. Entrenchment by a stream channel may be caused by:
   - (a) Sedimentation
   - (b) Change in base level
   - (c) Change in climate
   - (d) Change in base level or climate

   (Hint: See the first edition Slide Set.)

4. As a river moves from topographically high areas down to a broad floodplain and then to the ocean, the particles of sediment it carries:
   - (a) Become progressively coarser
   - (b) Become progressively finer
   - (c) Remain about the same size

   (Hint: See Animation 13.1.)

Web Project: The Mississippi Flood of 1993 was a truly awesome natural event. Not only was there was record flooding on the Mississippi River basin, but the Cedar River, Des Moines River, and Iowa River basins also had record floods—all at same time. The Army Corps of Engineers said that "these coincident events are a 'never-never' situation by Corps' standards." What the Corps usually designs for is the record event on one river basin. The problem here was that there were several record-setting events at one time spanning many river basins.

If you're interested in floods and how we attempt to mitigate their effects, a good link to check out is the <u>Mississippi River Field Trip</u> link (reached from the Chapter 13 page). This site describes a field trip taken by students in the Natural Hazard Mitigation course at Michigan Technological University to investigate the effects of the 1993 Mississippi River flood on communities along the river. The site features maps, interviews, and a report written by the students.

FINAL REVIEW

After reading this chapter, you should:

- Know how waters flow and how they collect and carry sediment
- Understand how stream valleys, channels, meanders, levees, and floodplains are formed
- Understand the significance of a stream's longitudinal profile
- Know the types and characteristics of drainage networks
- Understand the formation of deltas

ANSWER KEY

MEANDERS

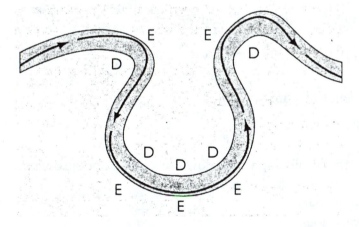

DRAINAGE PATTERNS

Dendritic: D, G, E, A, N   Trellis: F, C, J, K
Rectangular: I, F, M   Radial: H, B, J, L

PRACTICE MULTIPLE-CHOICE QUESTIONS

1. a, 2. c, 3. a, 4. b, 5. c, 6. d, 7. c, 8. c, 9. b, 10. d, 11. b, 12. b, 13. a, 14. a, 15. d, 16. c, 17. d, 18. b

Winds and Deserts

Wind is a potent force shaping the surface of the land, particularly in deserts. This chapter examines the properties of air flows; the wind as a major agent of erosion, transport, and deposition; and desert environments.

♦   Wind is a flow of air, and the rules of fluid flow that apply to water also apply to air flows. The main difference is that air flows are unconfined by solid boundaries. The extremely low density and viscosity of air make it turbulent even at low velocity. Although there are short-term variations in wind velocities and directions, a given location on Earth tends to have prevailing winds because of global atmospheric circulation (see Figure 14.1).

♦   Just as water transports sedimentary particles in a stream, the wind can move material across the surface of the Earth. The wind is most effective as a transportation agent in arid climates, where there is no moisture holding grains together. Because air has such low density, it does not usually move grains larger than sand.

Sand moves in the wind by sliding and rolling along the surface and by saltation, the jumping motion that temporarily suspends grains in a current of water or air. Saltation in air flows works the same way it does in a river, except that saltation is much more pronounced in air, because air is less viscous than water and because the impact of falling grains as they hit the surface induces higher jumps.

When the wind moves sand along a bed, ripples and dunes are formed. Ripples in sand, like those under water, are perpendicular to the current. Small ripples form first and become larger as wind speed increases. Ripples migrate in the direction of the wind over the backs of larger dunes.

Windblown sand usually consists of quartz grains, because quartz is so abundant. Windblown quartz grains often have a frosted or matte surface results from slow, long-continued dissolution by dew. Windblown calcium carbonate grains accumulate where there are abundant fragments of shells and coral. Dust includes microscopic rock and mineral fragments of all kinds, especially silicates.

◆   Deflation occurs when winds blow over an arid area, such as the desert floor, and remove the looser, finer sediment, leaving the larger gravels and pebbles. The surface left behind is called desert pavement, and the majority of desert floors worldwide consist of desert pavement.

As sand grains are blown along the surface, they can sandblast whatever happens to be exposed. Ventifacts are pebbles with several curved or almost flat surfaces that meet at sharp ridges. Each surface or facet is made by sandblasting of the pebble's windward side. Wind erosion can also produce yardangs, streamlined parallel ridges aligned with the direction of a strong prevailing wind.

◆   When the wind dies down, the coarser material it has carried is deposited in sand dunes. The formation of a sand dune results from some obstacle breaking the flow of the wind along the surface. Streamlines of wind separate around the obstacle and rejoin downwind, creating a wind-shadow zone in the lee of the obstacle. Sand grains blown into the shadow settle there and accumulate as a sand drift, then grow into a dune. As a dune grows, the combined movements of individual grains eventually cause the whole mound to migrate downwind.

Dune types vary with the amount of sand present and with the direction, duration, and strength of the wind. *(See the accompanying article on dune types.)*

In addition to sand grains, wind can deposit finer silts and clays. Such deposits, called loess, cover a huge amount of Earth's land surface.

◆   Deserts, which are defined as areas that receive less than 25 mm of rainfall per year, cover roughly 20 percent of Earth's land surface. Deserts may lie close to the equator; in mid-latitude regions where rainfall is low because moisture-laden winds either are blocked by mountain ranges or must travel great distances to the ocean; or in cold, dry polar regions. The location of deserts can be traced in large part to plate tectonics. *(See the accompanying article on deserts and plate tectonics.)*

◆   In the desert, physical weathering predominates over chemical weathering because of the lack of water required for chemical weathering. Desert varnish, however, is evidence that chemical weathering does occur, albeit at a much slower rate than in more humid climates. It is thought to form very slowly from a combination of dew, chemical weathering, and the sticking of windblown dust to exposed rock surfaces.

SAND DUNE TYPES

There are four main types of dunes (Figure 14.17): barchans, transverse dunes, blowouts, and linear dunes. A barchan is a crescent-shaped dune with the points of the crescent directed downwind. The slip face is the concave curve advancing downwind. Barchans are the products of limited sand supply and unidirectional winds. Several barchans may coalesce to form irregular ridges whose long direction lies transverse (perpendicular) to the wind direction. These transverse dunes are long, wavy ridges oriented at right angles to the prevailing unidirectional wind. Transverse dunes form in arid regions where abundant sand dominates the landscape and vegetation is absent. If a section of such a dune belt is deflated and stabilizing vegetation is overwhelmed by sand, a parabola-shaped dune called a blowout will form. In contrast to a barchan, which it superficially resembles, the arms of a blowout are directed upwind, and the slip face forms the convex curve advancing downwind. Linear dunes are long, straight ridges more or less parallel to the general direction of the prevailing winds. Most areas covered by linear dunes have a moderate sand supply, a rough pavement, and winds that are always in the same general direction.

DESERTS AND PLATE TECTONICS

In a way, deserts are reflections of plate tectonics. The mountains that create rain shadows are uplifted by collisions between converging continental and oceanic plates. The great distance separating central Asia from the oceans is a consequence of the size of the continent, a huge landmass assembled from smaller plates by continental drift. Large deserts are found at low latitudes because continental drift powered by plate tectonics has moved them there from higher latitudes. If the North American continent were to move south by more than 2000 km, the northern Great Plains of the United States and Canada would be converted to a hot, dry desert. Something like that has happened to Australia. Australia was once far to the south of its present position, and its interior had a moist, humid climate. Since then, Australia has moved northward into an arid subtropical zone, where its interior has become a desert.

Most desert streams flow only intermittently, but they account for most of the erosion in the desert when they do flow. Runoff is rapid and may cause flash floods along dry valley floors. Flash floods are often choked with loose weathering debris, and this sediment load moving at flood velocities abrades bedrock valleys.

♦   As sediment-laden flash floods dry up, they leave distinctive deposits on the floors of desert valleys. A flat fill of coarse debris often covers the entire valley floor, without the ordinary differentiation into channel, levees, and floodplains. Large alluvial fans form at mountain fronts in deserts because desert streams deposit much of their high sediment load on the fans.

Playa lakes are permanent or temporary lakes that accumulate in arid mountain valleys or basins. As the lake waters evaporate, weathering products are gradually precipitated. If evaporation is complete, the lakes become playas, flat beds of clay that are sometimes encrusted with precipitated salts.

♦   The desert is one of the most varied landscapes on Earth. Desert valleys have the same range of profiles as valleys elsewhere, but far more of them have steep walls caused by rapid erosion from mass movements and streams. Much of the landscape of deserts is shaped by rivers, but the valleys—called dry washes or wadis—are usually dry.

Pediments, broad, gently sloping platforms of bedrock left behind as a mountain front erodes and retreats from its valley, are characteristic desert landforms. Long-continued desert erosion also produces mesas—tablelike uplands capped by erosion-resistant beds bounded by steep-sided erosional cliffs.

The desert landscapes of millions of years ago have been obliterated by later erosion, but the preservation of desert sediments in the rock record has allowed geologists to reconstruct the time and extent of ancient deserts. The deserts of the geologic past were like those of today in all essential respects, so we know that arid climates and their consequences have dominated large sections of the continents for most of geologic time.

CHAPTER 14 OUTLINE

EROSIONAL AND DEPOSITIONAL FEATURES

Place an E next to an item if it is an erosional feature, a D if it is a depositional feature.

1. _____ Dune   6. _____ Yardang

2. _____ Playa   7. _____ Barchan

3. _____ Mesa   8. _____ Dry wash

4. _____ Pediment   9. _____ Ventifact

5. _____ Alluvial fan   10. _____ Loess

DUNE TYPES AND PREVAILING WINDS

Give the names of the dune types shown below, then draw arrows on the figures to indicate the direction(s) of the prevailing winds.

A. _____

B. _____

C. _____

D. _____

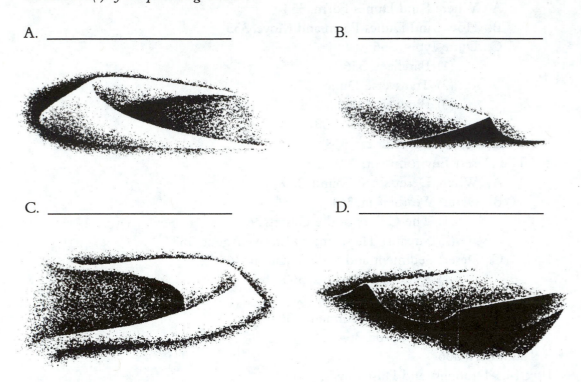

PRACTICE MULTIPLE-CHOICE QUESTIONS

Circle the option that best answers the question.

1. Wind cannot carry large particles for great distances because:
    (a) Air currents are unconfined by boundaries
    (b) Air has a very low density
    (c) Particles are never lifted into the air
    (d) Air flows are turbulent

2. Which of the following ways in which particles move plays the largest role in wind transport?
    (a) Rolling
    (b) Suspension
    (c) Saltation
    (d) Solution

3. Desert pavement is produced by:
    (a) Deflation of the finer sediments by the wind
    (b) Development of a surface crust of clay and iron oxides due to drying out of the surface
    (c) Headward erosion by streams
    (d) Deposition of gravel and pebbles by flash floods

4. Physical weathering by sandblasting is only effective when:
    (a) Wind velocities are above 50 km per hour
    (b) Air turbulence is maximized because of obstacles in the path of wind flow
    (c) Particles of sufficient hardness are blown along near the ground
    (d) Rivers supply sand on a periodic basis

5. Dunes formed by the wind are composed of:
    (a) Calcite and quartz
    (b) Only soft minerals such as calcite and gypsum
    (c) Only quartz
    (d) Any wind-driven material

6. Erosion of a sand dune occurs:
    (a) On the slip face of the dune
    (b) When dry grains are carried away by the wind
    (c) When grains saltate into the wind
    (d) Only along the points of a dune

7. The feature found in a sand dune that is related to the angle of repose of sand grains is the:
    (a) Windward slope
    (b) Slip face
    (c) Height of its crest as measured in relation to the width of the dune
    (d) Width of the dune as measured on the desert floor perpendicular to its migration direction

8. Barchan dunes are recognized by the fact that:
    (a) Their tips point downwind
    (b) Their tips point upwind
    (c) They form ridges parallel to the prevailing wind direction
    (d) Their slip faces are always pointed into the wind

9. A sand erg is:
    (a) A measure of the amount of energy required to move the sand
    (b) A term used in the Middle East to describe barchan dunes
    (c) An extensive area covered by sand dunes; a sea of sand
    (d) A term used to measure the depth of blowout

10. Sediment known as loess is:
    (a) Dust transported by wind in suspension and deposited on land
    (b) Dust transported by wind, deposited, and subsequently eroded and re-deposited by water
    (c) Fine-grained sediment transported and deposited by saltation
    (d) Fine-grained silty sediment deposited in desert lakes

11. Rain shadow deserts are:
    (a) Located within 10 degrees of the equator
    (b) Confined to the middle of large continents
    (c) Found within 100 km of coastal regions
    (d) Located on the lee side of mountain ranges

CD RESOURCES

SUPPLEMENTAL DIAGRAMS

14.27 Saltation layers
14.28 Loess and other eolian deposits in central United States

SUPPLEMENTAL PHOTOS

14.29 Ripplemarks in gypsum sand
14.30 Interfering sets of cross-beds in an ancient eolian sandstone
14.31 Desert pavement
14.32 Quartzite rock smoothed and polished by sandblasting
14.33 Basalt ventifact
14.34 Looking down the slip face of a dune
14.35 Windward and leeward dune slopes
14.36 Barchan dunes
14.37 Transverse dunes on the Saudi Arabian desert
14.38 Large wavelike transverse dunes
14.39 Blowout dunes, Imperial County, California
14.40 Blowout dunes, south shore of Lake Michigan
14.41 Draas in the Saudi Arabian desert
14.42 Cliffs of loess, Tennessee
14.43 Salt polygons in Death Valley playa
14.44 Mountain pediments and playa
14.45 Mountain front with pediments in the distance
14.46 Modern mudcracks in an old pond

FIRST EDITION SLIDE SET

Slide 14.1 Gran Desierto, northern Sonora, Mexico
Slide 14.2 Star dunes, Gran Desierto
Slide 14.3 Linear dunes, Simpson Desert, Australia
Slide 14.4 Barchan dunes, Algondones, near Yuma, Arizona
Slide 14.5 Sonoran Desert, Kofa Mountains, western Arizona
Slide 14.6 Desert pavement and desert varnish, Kofa Mountains, western Arizona
Slide 14.7 Ventifacts, Namib Desert, southwestern Africa
Slide 14.8 Dust storm, Niger, Africa

ILLUSTRATED GLOSSARY

EXERCISES

Major Deserts of the World

ANIMATIONS

14.1   Winds Creating Sand Dunes

CD LAB

1. Which of the following states in the central United States have loess deposits more than 25 feet deep?
    (a) Tennessee, Illinois, Iowa
    (b) Mississippi, Louisiana, Kentucky
    (c) Michigan, Illinois, Indiana
    (d) Nebraska, Kansas, Iowa

    (Hint: See the Chapter 14 supplemental diagrams.)

2. What is the predominant composition of the sand at White Sands National Monument, New Mexico?
    (a) Quartz
    (b) Calcite
    (c) Gypsum
    (d) Quartzite

    (Hint: See the Chapter 14 supplemental photos.)

3. Complex patterns of interfering cross-beds in eolian sandstones are a result of:
    (a) Meandering desert streams
    (b) Changing wind directions
    (c) Differential weathering and erosion
    (d) Alluvial fans

    (Hint: See the Chapter 14 supplemental photos.)

4. The largest active sand sea in North America is the:
    (a) Gran Desierto
    (b) Mojave Desert
    (c) Sonora Desert
    (d) Sahara Desert

    (Hint: See the first edition Slide Set.)

5. Barchanoid dunes are transitional between:
    (a) Linear and barchan dunes
    (b) Transverse and barchan dunes
    (c) Barchan and blowout dunes
    (d) Barchan and linear dunes

    (Hint: See Animation 14.1.)

FINAL REVIEW

After reading this chapter, you should:

- Understand how winds transport sand
- Know what deflation and sandblasting are
- Understand how winds deposit sediment
- Know the different forms of wind deposits
- Know how sand dunes form and move
- Know how deserts form
- Understand how the desert environment works

ANSWER KEY

EROSIONAL AND DEPOSITIONAL FEATURES

1. D, 2. D, 3. E, 4. E, 5. D, 6. E, 7. D, 8. E, 9. E, 10. D

DUNE TYPES AND PREVAILING WINDS

A. Barchan, B. Transverse, C. Blowout, D. Linear
The wind directions are shown in Figure 14.17 on page 357 of the text.

PRACTICE MULTIPLE-CHOICE QUESTIONS

1. b, 2. c, 3. a, 4. c, 5. d, 6. b, 7. b, 8. a, 9. c, 10. a, 11. d

15

Glaciers: The Work of Ice

Much of the water on Earth's surface is frozen in glaciers, and the action of glaciers creates some of the most spectacular scenery on Earth. This chapter examines how glaciers form; how they move; and how they erode, transport, and deposit sediment. It also discusses how ice ages come and go with changes in global climate

♦ Geologically speaking, a block of ice is a rock composed of crystalline grains of the mineral ice. Although ice shares many characteristics of rocks, it is much less dense than most rocks and has an extremely low melting temperature.

Glaciers are large masses of ice on land that show evidence of being in motion or of once having moved. There are two basic types of glaciers: valley glaciers and continental glaciers. Valley glaciers form high in mountain ranges and flow down the bedrock valleys. A continental glacier is an extremely slow moving, thick sheet of ice, much larger than a valley glacier. Ice sheets cover much of Greenland and Antarctica.

Ice caps are the masses of ice formed at Earth's North and South Poles. Most of the Arctic ice cap lies over water and is generally not referred to as a glacier. Because almost all of the Antarctic ice cap lies over land, it is considered to be a continental glacier.

♦ Glaciers form where it is cold enough for snow to remain on the ground year-round: at high latitudes and at high altitudes. The formation of snow and glaciers requires moisture as well as cold. The dry leeward side of a mountain range is likely not to have glaciers.

The transformation of loose snow into glacial ice is an important step in the creation of a glacier. The continual buildup of snow creates increased pressure on the snow that is buried, compacting it into firn. Further burial and aging produce solid glacial ice as the smallest grains recrystallize and cement all the grains together. Snow can be thought of as a sediment that is transformed by burial into a metamorphic rock called ice.

♦   A typical glacier grows slightly during the winter, as snow falls on the glacial surface and is converted to ice. The amount of snow added to the glacier annually is called its accumulation. After enough ice has accumulated, the mass will begin to flow downhill under the force of gravity.

The total amount of ice that a glacier loses each year is called ablation. Ice can be lost by melting, iceberg calving, sublimation, or wind erosion. Glaciers lose ice mainly by melting and calving.

A glacier will grow or shrink depending on whether accumulation exceeds or is less than ablation. The net gain or loss of ice is called the glacial budget. When accumulation equals ablation over a long period, the glacier remains of constant size. If accumulation exceeds ablation, the glacier grows; if ablation exceeds accumulation, the glacier shrinks.

♦   A tremendous quantity of fresh water is tied up in glacial ice, making glaciers a possible source of fresh water for human use. Some geologists have even suggested that icebergs might be towed to water-poor regions in the future.

♦   When ice becomes thick enough for gravity to overcome the ice's resistance to movement, it starts to move and thus becomes a glacier. Glaciers move in a laminar flow, by two main mechanisms: plastic flow and basal slip. In plastic flow, the ice deforms and slides internally on a microscopic scale. In basal slip, the ice, lubricated by meltwater, slides downslope along the base of a glacier, like a brick sliding down an inclined board.

Plastic flow dominates in very cold regions, where the ice throughout the glacier is well below the freezing point. The basal ice is frozen to the ground, and most of the movement of these cold, dry glaciers takes place above the base by plastic deformation. What little movement there is rips up any detachable pieces of bedrock or soil. Basal slip occurs in warmer areas where the ice at the base can slide along on meltwater.

The upper parts of glaciers have little pressure on them from overlying ice, and the ice here behaves as a brittle solid, cracking as it is dragged along by the plastic flow of the ice below. These cracks, called crevasses, break up the surface ice into many blocks. They tend to occur where the ice drags against bedrock walls, at curves in the valley, and where the slope steepens sharply.

A glacier's overall rate of motion varies with the amount of ice present and the slope of the surface on which it lies. A typical valley glacier may move at up to 75 m per year, whereas continental glaciers move more slowly, perhaps only 8 or 9 m per year.

◆   Glaciers have an amazing capacity to erode solid rock. A valley glacier only a few hundred meters wide can tear up and crush millions of tons of bedrock in a single year. At its base and sides, a glacier engulfs jointed blocks and breaks and grinds them against the rock pavement below, producing fragments from house-sized boulders to fine material called rock flour. As a glacier drags rocks along its base, they scratch the pavement as they grind against it, forming striations. Small hills of bedrock called roches moutonées are smoothed by the ice on their upcurrent side and plucked by the ice to a rough, steep slope on the downcurrent side.

A valley glacier carves a series of erosional forms as it flows from its origin to its lower edge. At the head of the glacial valley, the glacier carves out a cirque, a hollow shaped like one half of an inverted cone. With continued erosion, cirques at the heads of adjacent valleys gradually meet at the mountaintops, producing sharp crests called aretes along the divide. As a valley glacier moves down from its cirque, it excavates a valley or deepens a preexisting river valley, creating a characteristic U-shaped valley. Glacial erosion can also create a hanging valley, a tributary valley whose floor lies high above the main valley floor.

Valley glaciers at coastlines may erode their valley floors below sea level. When the ice retreats, these steep-walled valleys are flooded with seawater, producing glaciated arms of the sea called fjords.

◆   Glacial erosion creates an enormous amount of debris, and ice transports this debris to the edge of the glacier, where it is deposited or carried away by meltwater streams. Ice transports debris very effectively because the material it picks up does not settle out, as does the load carried by a river. As a current, ice has an extremely high competence and carrying capacity.

When glacial ice melts, it deposits a poorly sorted, heterogeneous load of sediment called till. A wide range of particle sizes differentiates glacial sediment from the sorted material deposited by streams and winds. The large boulders deposited by ice are called erratics because their composition is very different from that of local rocks.

Meltwater streams flowing in tunnels within and beneath the ice and in streams at the ice front may pick up, transport, and deposit some of the material. This outwash is stratified and well sorted and may be cross-bedded.

An accumulation of material carried by ice or deposited as till is called a moraine. There are many kinds of moraines, each named for its position with respect to the glacier that formed it: end moraines, terminal moraines, lateral moraines, medial moraines, and ground moraines.

Drumlins are large, streamlined hills of till and bedrock that parallel the direction of ice movement. Their origin is not completely understood.

Deposits of outwash take a variety of forms. Kames are small hills of sand and gravel dumped at the edge of the ice. Silts and clays are deposited on the bottom of a lake at the edge of the ice in a series of alternating coarse and fine layers called varves. Eskers are long, narrow, winding ridges of sand and gravel found in the middle of ground moraines. Kettles are hollows that often are steep-sided and may be occupied by ponds or lakes.

♦   Permafrost, perennially frozen soil, occurs in very cold regions where the temperature never rises enough to melt more than a thin surface layer of ice. Permafrost covers as much as 25 percent of Earth's total land area. The thickness of the permafrost layer varies from region to region.

♦   Because snow and prolonged cold are necessary for the formation of glaciers, glacial sediments and erosional forms are indicators of past and present climates. The former large extent of glaciers is evidence of a recent past in which the climate over large regions was much colder than it is today. At some times much farther back in the geologic past, glaciers covered regions that are now tropical jungles. At other times, parts of the globe now covered with ice were warm and humid, and there were no polar ice caps. The former extent of glaciers can be used both to trace climates of the past and to predict future changes in climate. *(See the accompanying article on ice ages.)*

ICE AGES

When the Swiss geologist Louis Agassiz found a variety of deposits in North America that resembled the glacial deposits he had seen in the Swiss Alps, he concluded that much of the northern part of the continent had been covered by ice sheets in the recent geologic past. The areas covered by these glacial deposits were so vast that the ice that produced them must have been a continental glacier larger than Greenland or Antarctica. Later information revealed that the last ice had disappeared 10,000 years ago, at the end of the Pleistocene epoch.

It soon became clear that there were multiple glacial ages during the Pleistocene, with warmer interglacial intervals between them. This was evidenced by the presence of several layers of drift, alternating with well-developed soils containing fossils of warm-climate plants. Similar evidence was found in ocean sediments that revealed

fluctuations in the temperature of the ocean over the same period. There was also evidence of periods when evaporation removed more water from the ocean than was replaced by runoff. Evidently, the water was tied up in continental ice during the glacial periods. During the maximum extent of the most recent glaciation, 18,000 years ago, sea level dropped about 130 m, corresponding to about 70 million cubic kilometers of ice, or almost three times the amount of ice on Earth today. The drop in sea level changed the shapes of the continents as the retreating sea exposed portions of their continental shelves.

The causes of climate change and glacial periods have been debated ever since geologists learned about the ice ages. What could cause cold periods to alternate with relatively short warm periods over the past 2 million years? The alternation between glacial and interglacial ages is best explained in terms of astronomical cycles. We know that Earth's orbit varies, so that the distance between Earth and the Sun is not constant. Earth also wobbles slightly as it rotates on its axis, so that it periodically tilts toward and then away from the Sun. A combination of these two variables could produce a warming and cooling of the Earth on a 100,000-year cycle.

We also must account for the fact that the Pleistocene glaciation was not unique in Earth's history. We know from glacial striations and lithified ancient tills, called tillites, that glaciations covered parts of the continents many times in the geologic past, long before the Pleistocene. At present, the best explanation is based on the motion of Earth's tectonic plates. As the plates move, they interfere with the circulation of water in the ocean. Ocean water distributes a great deal of Earth's heat. When continents are arranged as they are today, they restrict the flow of warm water from the equator to the poles, resulting in a cooling of the polar areas and ice accumulation there.

CHAPTER 15 OUTLINE

EROSIONAL AND SEDIMENTARY LANDFORMS

Place an E next to an item if it is an erosional landform, an S if it is a sedimentary landform.

1.	_____ Kame		6.	_____ Arete
2.	_____ Drumlin		7.	_____ Moraine
3.	_____ Roche moutonée		8.	_____ Hanging valley
4.	_____ Kettle		9.	_____ Cirque
5.	_____ Esker		10.	_____ Fjord

PRACTICE MULTIPLE-CHOICE QUESTIONS

Circle the option that best answers the question.

1. Approximately what percentage of the Earth's surface is covered by ice?
   (a) 10
   (b) 25
   (c) 40
   (d) 50

2. Calving refers to the:
    (a) Amount of terminal moraine deposited at the toe of a glacier
    (b) Downhill movement of glaciers
    (c) Removal of ice due to melting or sublimation
    (d) Breaking off of large pieces of ice where the glacier terminates along a shoreline

3. Cracks in the surface ice of a glacier are called:
    (a) Drift
    (b) Crevasses
    (c) Striations
    (d) Glacial polish

4. The velocity of the movement of ice in a valley glacier will be higher:
    (a) At the center of the glacier, just below its surface
    (b) At the edges of the glacier, near its contact with the bedrock
    (c) At the leading edge of the glacier
    (d) Equally throughout its cross-sectional area

5. Glacial striations are formed by:
    (a) A glacier as it retreats
    (b) Pieces of rock frozen into the bottom of a glacier
    (c) Very fine silt and sand that may be incorporated into the bottom of glacial ice
    (d) Deposition of glacial till

6. An arete is:
    (a) A depositional feature found on glaciated terrain
    (b) Another term for a pass or low point between peaks
    (c) A sharp, jagged ridge formed by glaciers eroding a ridge from both sides
    (d) A steep-walled bowl at the head of a glacier

7. The cross-sectional shape of most glaciated valleys is:
    (a) Box-shaped (steep walls, flat bottom)
    (b) V-shaped
    (c) U-shaped
    (d) Unpredictable because of differences in bedrock types

8. Hanging valleys are most commonly associated with:
    (a) River erosion
    (b) Wind erosion
    (c) Continental glaciation
    (d) Valley glaciation

9. Fjords are:
    (a) A string of cirques
    (b) Composed primarily of till deposits
    (c) The result of submarine erosion by glacial ice
    (d) Flooded glacial troughs along shorelines

10. An erratic is:
    (a) A glacially deposited rock whose composition differs from that of the bedrock beneath it
    (b) A mound of heterogeneous sediment
    (c) A feature associated with kettles
    (d) Another name for a roche moutonée

11. An accumulation of material carried or deposited by ice is called:
    (a) Till
    (b) Drift
    (c) A moraine
    (d) A drumlin

12. A drumlin is:
    (a) A low, streamlined hill constructed of glacial till
    (b) A block of bedrock that was not plucked away by the glacier
    (c) A closed depression formed when a block of ice melts out of water-deposited glacial drift
    (d) A sinuous ridge of water-deposited glacial debris

13. A kettle is a depression:
    (a) Created by the erosive action of glacial ice
    (b) That results from the melting of ice blocks left behind by a retreating continental glacier
    (c) That is formed in relatively soft rock
    (d) That often has smooth, shallow sides

14. Which of the following glacial landforms is not the product of deposition by water?
    (a) Drumlins
    (b) Kames
    (c) Eskers
    (d) Outwash plains

15. Which of the following cannot be used to determine the direction of ice movement?

  (a) Distribution of rock types
  (b) Striations
  (c) Varves
  (d) Drumlins

16. The maximum extent of the most recent glaciation occurred:

  (a) 10,000 years ago
  (b) 18,000 years ago
  (c) 100,000 years ago
  (d) About 2 million years ago

CD Resources

SUPPLEMENTAL DIAGRAMS

15.37 Comparison of glacial till and outwash
15.38 Comparison of roche moutonée and drumlin
15.39 Glacial geology of the contiguous United States
15.40 Late Paleozoic glaciated areas of Gondwanaland
15.41 Locations of late Paleozoic glaciation and directions of flow

SUPPLEMENTAL PHOTOS

15.42 Malaspina Glacier, Alaska (aerial view)
15.43 Hubbard Glacier, Alaska (aerial view)
15.44 Meltwater at a glacier terminus
15.45 Glacier terminus
15.46 Retreating glacier with trimline, Saskatchewan, Canada
15.47 Glacial till on striated sandstone
15.48 Glacial erratic in Central Park, New York City
15.49 Medial moraine
15.50 Ground moraine left by a valley glacier
15.51 Kettle at end of large esker
15.52 Patterned ground, Barrow district, Alaska
15.53 Differential subsidence of railroad bed caused by permafrost thaw
15.54 Gravel road with severe differential subsidence

FIRST EDITION SLIDE SET

Slide 15.1 Taku and Norris glaciers, Juneau Icefield, Alaska
Slide 15.2 Ruth Glacier, Alaska Range
Slide 15.3 The Great Gorge, Ruth Glacier, Alaska Range
Slide 15.4 Half Dome and Tenaya Canyon, Yosemite, California

Slide 15.5 Lateral moraines, McGee Creek, Owens Valley, California

Slide 15.6 Glacial erratic, South Bubble, Acadia, Maine

Slide 15.7 Glacial striations and polish, Athabasca Glacier, Columbia Icefield, Canada

Slide 15.8 Lateral moraines, Victoria Glacier, Lake Louise, Alberta, Canada

Slide 15.9 McCarty Fjord, Kenai Peninsula, Alaska

Slide 15.10 Athabasca Glacier, Columbia Icefield, Canada

SECOND EDITION SLIDE SET

Slide 15.1 Deposits and features at the front of a retreating ice sheet, Iceland

Slide 15.2 Russian icebreaker at the North Pole, Arctic Ocean

Slide 15.3 Open water of the Arctic Ocean at the North Pole

Slide 15.4 Flow of sea ice in Arctic Ocean mimics the movement of Earth's crustal plates

Slide 15.5 Queen Maud Mountains in Antarctica

Slide 15.6 Laboratory studies of ice core from Antarctica

Slide 15.7 Snow accumulation study on the Lemon Glacier, Juneau Icefield, Alaska

EXPANSION MODULES

Chapter 25: The Geology of Canada

Chapter 35: The Rising Seas (*Scientific American* article)

ANIMATIONS

15.1   Glaciers: Ablation

15.2   Glaciers: Accumulation

15.3   Glaciers: Equilibrium

ILLUSTRATED GLOSSARY

EXERCISES

Glacial Landscapes

CD LAB

1. The oldest known glacial deposits are of _____ age:
   (a) Archean
   (b) Proterozoic
   (c) Paleozoic
   (d) Pleistocene

   (Hint: See the Geologic Time Scale tool.)

2. The deepest gorge in North America is the:
    (a) Grand Canyon, Arizona
    (b) Snake River Canyon, Idaho
    (c) Great Gorge, Alaska
    (d) Tenaya Canyon, Yosemite, California

    (Hint: See the first edition Slide Set.)

3. During the twentieth century, most tidewater glaciers in Alaska have undergone:
    (a) Rapid expansion
    (b) Slow expansion
    (c) Rapid retreat
    (d) Slow retreat

    (Hint: See the first edition Slide Set.)

4. The flow of sea ice in the Arctic Ocean mimics:
    (a) The movement of tectonic plates
    (b) A meandering river
    (c) Iceberg calving
    (d) Isostasy

    (Hint: See the second edition Slide Set.)

5. Canada has experienced at least _____ major phases of glaciation during the Quarternary period:
    (a) Three
    (b) Four
    (c) Five
    (d) Seven

    (Hint: See Expansion Module Chapter 25.)

6. Which of the following factors arising from global warming is most likely to cause a calamitous rise in sea level?
    (a) Thermal expansion of seawater as the water temperature rises
    (b) Melting of water tied up in valley glaciers
    (c) Melting of the Antarctic ice cap

    (Hint: See Expansion Module Chapter 35.)

Web Project: The Chapter 15 page contains links to several interesting sites, including Glaciers, Glacial Geology Photo Gallery, and Glacial Geology at the University of Cincinnati (all of which have extensive collections of glacier photos); Global Warming Update; and Worthington Glacier, Alaska (a research project on the mechanics of glaciers). The National Snow and Ice Data Center has numerous photos, maps, and educational resources available.

FINAL REVIEW

After reading this chapter, you should:

- Understand what ice is in geological terms
- Know how glaciers form, grow, and shrink
- Understand how glaciers move
- Know the effects of glacial erosion and the landforms it produces
- Understand glacial sedimentation and know the types of sedimentary landforms
- Understand the effects of the ice ages

ANSWER KEY

EROSIONAL AND SEDIMENTARY LANDFORMS

1. S, 2. S, 3. E, 4. S, 5. S, 6. E, 7. S, 8. E, 9. E, 10. E

PRACTICE MULTIPLE-CHOICE QUESTIONS

1. a, 2. d, 3. b, 4. a, 5. b, 6. c, 7. c, 8. d, 9. d, 10. a, 11. c, 12. a, 13. b, 14. a, 15. c, 16. b

16 Landscape Evolution

Earth's landscape is made up of a great variety of landforms. How do such varied landscapes form in different locales? This chapter examines the ways that major geologic forces—geologic history, tectonics, climate, weathering, erosion, and sedimentation—interact in the dynamic process that carves the landscape.

◆   To construct accurate maps of Earth's surface, we need to know the topography, or the varying heights above sea level. We then express the altitude, or vertical distance above sea level, as elevation. A topographic map shows the distribution of elevations in an area and usually represents this distribution by contours, lines that connect points of equal elevation.

Relief is the difference between the highest and the lowest elevations in a particular area and is a measure of the roughness of a terrain. The higher the relief, the more rugged the topography. Most regions of high elevation also have high relief, and most areas of low elevation have low relief.

◆   Landforms are a region's characteristic landscape features created by uplift, erosion, and sedimentation. Landforms are guides to the region's geologic structure and history.

A mountain is a large mass of rock that projects well above its surroundings and is usually found with others in ranges. The difference between a mountain and a hill is one of size compared to the surrounding area. In general, landforms more than several hundred meters above their surroundings are called mountains.

A plateau is a large, broad, flat area of appreciable elevation above the neighboring terrain. In the western United States, a small, flat elevation with steep slopes on all sides is called a mesa. Plateaus form where tectonic activity produces a general uplift.

Cuestas are asymmetrical ridges in a tilted and eroded series of beds of alternating weak and strong resistance to erosion. Hogbacks are steep, narrow, symmetrical ridges that form in steeply dipping or vertical erosion-resistant beds.

In young mountains, during the early stages of folding and uplift, the upfolds (anticlines) form ridges and the downfolds (synclines) form valleys. As uplift slows and erosion increases, the anticlines may form valleys and the synclines ridges. This happens where the rocks exert strong control on topography by their variable resistance to erosion.

The specific geometry of river valleys varies in regions of different general topography and bedrock type, ranging from narrow gorges in mountain belts and erosion-resistant rocks to wide, shallow valleys in plains and easily eroded rocks. A badland is a deeply gullied topography resulting from the fast erosion of easily erodible shales and clays.

Tectonic valleys are long, narrow, flat-floored, and bounded on one or both sides by faults. This shape results from downward movement of the valley caused by crustal subsidence along the faults. A rift valley is a special type of tectonic valley formed by plate divergence. Basins form where Earth's surface has subsided as a result of tectonic movement, often between mountain ranges.

♦   Landforms are primarily the products of erosion, transportation, and sedimentation, but tectonics is also a factor. The profile of a river valley in tectonically elevated high mountains differs from that of one in tectonically stable low plains. Bedrock lithology also plays a role: river valleys that cut through easily erodible sediments and rocks are broader and have gentler slopes than those that pass through erosion-resistant rocks. Climate also strongly influences landscape: a hot, dry desert landscape is very different from a polar landscape. The topography of a region strongly influences the rate of weathering and erosion. All these factors interact to produce landforms.

Tectonics, driven by plate motions, elevates mountains and lowers tectonic valleys and basins. Erosion lowers the land surface and carves bedrock into valleys and slopes. The interaction between the two is a negative-feedback process in which one action produces an effect (the feedback) that tends to slow the original action and stabilize the process at a lower rate. Tectonic uplift provokes an increase in erosion rate: the higher the mountains grow, the faster erosion wears them down. While mountain building continues, elevations stay high or increase. When mountain building slows, the mountains rise more slowly or stop rising entirely. As growth slows or stops, erosion starts to dominate and elevations begin to decrease. As the lowering of mountains proceeds, erosion also slows, and the whole process eventually tapers off. Elevation is a balance between uplift and erosion rate.

LANDSCAPE EVOLUTION AND PLATE TECTONICS

Current views of landscape evolution emphasize the balance between erosion and tectonic uplift. If uplift is faster, the mountains rise; if erosion is faster, the mountains are lowered.

What controls the rate and duration of tectonic activity? Uplift begins when two plates converge. The convergence of two oceanic plates leads to arcs of volcanic islands, whose topography is dominated by active volcanism, with erosion lagging. The convergence of an oceanic plate and a continental plate leads to a topography like that of the Pacific coast of South America, where the volcanic arc is on land. Extensive volcanism and lateral compression formed the Andes, while sedimentation and subsidence in the forearc region between the Andes and the Peru–Chile Trench offshore formed a narrow coastal plain. The convergence of two continental plates leads to high mountains, such as the Himalayas.

The rate of uplift varies with the speed of plate convergence. For reasons not fully understood, convergence may slow to a halt at times and then resume. Plate movements may also accelerate and then slow. High rates of convergence lead to rapid uplifts that outstrip erosion. Erosion dominates when convergence slows. Erosion may win out permanently if the geometry of plate motions changes and new plate boundaries are created far from the mountains.

Erosion may actually promote tectonic uplift. Changes in climate may increase weathering and erosion rates and thus strip great weights of rock from the surface. This may lead to a response of the interior that uplifts the surface and raises mountains.

Although there are broad categories of landscape, each region has its own geologic history. That history determined the rock types at the surface and their structures, both of which strongly affect the course of landscape evolution. The latitude of a region is also important, and a region may move from low to higher latitudes by continental drift.

Climate also interacts with topography. At higher elevations, temperature drops and rainfall is reduced, vegetation is sparse, and physical weathering predominates over chemical weathering. Another effect of topography on climate can be seen in the rain shadows formed on the leeward slopes of mountain ranges. The climate becomes cooler as latitude increases, and the corresponding effects on weathering and erosion are much the same as those caused by increasing elevation. The interactions of climate, topography, and latitude have strong effects on landscape by enhancing or suppressing chemical and physical weathering. Topography, which is controlled primarily by tectonics, is the single most dominant factor in landscape evolution. In all latitudes and climates, the slopes of high mountains are steep and rugged and lowlands tend to have gentler topographies.

♦   The landforms of North America show the dominance of tectonics but also suggest the importance of age. The highest and ruggedest mountains are the young western and northern ranges. The lower and older eastern mountains are rounded and gentle. The oldest terrain of all, the Precambrian Canadian Shield, is worn down to a low-lying plain. To take age into account in the formation of landscape, we need to know how landscapes evolve over geologic time.

♦   Landscape evolution is largely a result of the balance between erosion and tectonic uplift. *(See the accompanying article on landscape evolution and plate tectonics.)*

CHAPTER 16 OUTLINE

LANDSCAPE TERMS AND CONCEPTS

Fill in the blanks with the appropriate word or term that completes the sentence or statement.

1. Asymmetrical ridges produced by the different erosion rates of the rocks forming the ridges are called _____ and _____.

2. Deeply gullied surfaces that form because of rapid erosion of soft rocks such as shales and clays are called _____.

3. The formation of landscapes is eventually explained by forces related to _____.

4. In deserts, _____ weathering is more active than _____ weathering.

5. The height of a mountain range is controlled by a _____ _____ process.

6. The effects of latitude mimic those of _____.

7. Tectonics proposes, _____ disposes.

PRACTICE MULTIPLE-CHOICE QUESTIONS

Circle the option that best answers the question.

1. The topography of an area is a general description of its:
    (a) Overall surface shape
    (b) Greatest difference in elevations
    (c) Vertical distance above sea level
    (d) Rock types in relation to their degree of weathering

2. Elevation is a measure of:
    (a) Height above the lowest local point on land
    (b) Height above the lowest local point
    (c) Height above sea level
    (d) Height from the base of the landform to its highest point

3. Which of the following statements is true?
    (a) Areas of low elevation tend to have high relief
    (b) Areas of low relief tend to have high elevation
    (c) Areas of high elevation also tend to have high relief
    (d) Relief and elevation generally are not related

4. Although the term *mountain* does not have a strict definition, which characteristic best defines the term?
    (a) It is large mass of rock
    (b) It is always formed by tectonic forces
    (c) It is a landform having several hundred meters of relief
    (d) It is a geologically young feature

5. Which of the following landforms is the largest?
   (a) Mesa
   (b) Butte
   (c) Hogback
   (d) Plateau

6. The earliest stages of development of a river valley would have:
   (a) A simple V-shaped profile
   (b) A simple U-shaped profile
   (c) A low stream gradient
   (d) A well-established floodplain

7. Structurally controlled ridges and valleys are most often associated with:
   (a) Areas of high relief
   (b) Areas of low relief
   (c) Horizontal rocks that have been faulted
   (d) Horizontal rocks that have been folded

8. A rift valley is a special form of:
   (a) Tectonic valley produced by plate divergence
   (b) Tectonic valley produced by plate convergence
   (c) River valley produced by an abrupt change in grade, as at a mountain front
   (d) V-shaped river valley

9. Physical weathering will be the dominant process affecting the formation of the future landscape in:
   (a) Florida
   (b) Regions along the equator
   (c) The highest elevations of the Rocky Mountains
   (d) The Caribbean Islands

10. Which of the following statements is true:
    (a) Elevation is the result of a balance between tectonics and erosion
    (b) Elevation is primarily a result of tectonics
    (c) Elevation is primarily a result of erosion
    (d) Tectonics and erosion are unrelated

11. The one factor that can be considered the most dominant in all landscape evolution is:
    (a) Climate
    (b) Latitude
    (c) Topography
    (d) All three are essentially equal in importance

12. Which of the following sections of the United States displays the greatest relief?
    (a) The Rocky Mountains
    (b) The Midwest
    (c) The coast of New England
    (d) The coast of the Gulf of Mexico

13. The basins in western Nevada formed as a result of:
    (a) Severe erosion of the landscape following melting of the Pleistocene glaciers
    (b) Extensional tectonic movement that produced mountains and basins
    (c) Compressional forces that folded the upper crustal blocks
    (d) Subsidence related to volcanic activity

14. The mountain range that formed as the result of active plate subduction is the:
    (a) Himalayas in Nepal and Tibet
    (b) Andes along the west coast of South America
    (c) Appalachians in the eastern United States
    (d) Ozarks of central Arkansas

CD RESOURCES

SUPPLEMENTAL DIAGRAMS

16.18 Topography characteristic of younger and older mountains
16.19 Physiographic regions and provinces of the contiguous United States
16.20 Major geologic features of North America
16.21 Geologic provinces of North America late in Proterozoic time
16.22 Flooding of North America during Maastrichtian time

SUPPLEMENTAL PHOTOS

16.23 Hogback ridges

FIRST EDITION SLIDE SET

Slide 16.1 Landforms of the contiguous United States
Slide 16.2 Marble Canyon and the Grand Canyon, satellite image
Slide 16.3 Marble Canyon, aerial view

SECOND EDITION SLIDE SET

Refer to Slide 10.8 and the slides listed for Chapter 21.

EXPANSION MODULES

The expansion module chapter on Canada (Chapter 25) and the chapters on regions of the United States (Chapters 26–30) contain photos of landforms in these areas. Chapter 31, The Himalayas and Tibetan Plateau, has many excellent photos of the Himalayas.

ANIMATIONS

20.1   Plate Tectonics: Plate Divergence
20.2   Plate Tectonics: Continent-Continent Convergence
20.3   Plate Tectonics: Ocean–Continent Convergence

VIDEOS

26.1   America by Air: Mount Shasta, California
26.5   America by Air: Yosemite Valley, California
26.6   America by Air: Mount Whitney, California
27.1   America by Air: Canyonlands, Utah
27.2   America by Air: Great Plains
27.3   America by Air: Northern Rocky Mountains, Montana
29.1   America by Air: Everglades, Florida
29.2   America by Air: Valley and Ridge, Appalachians

ILLUSTRATED GLOSSARY

EXERCISES

Landform Provinces of North America

CD LAB

1. The Great Valley of California is a relic of a forearc basin associated with a subduction zone, which was active:
    (a) In the early Paleozoic
    (b) In the middle to late Paleozoic
    (c) From the early Mesozoic to the middle Cenozoic
    (d) In the Jurassic

    (Hint: See the first edition Slide Set.)

2. The rocks of the Sierra Nevada are largely the roots of a predominantly _____ volcanic arc.
    (a) Permian
    (b) Triassic
    (c) Jurassic
    (d) Cretaceous

    (Hint: See the first edition Slide Set.)

3. The north rim of the Grand Canyon is defined by:
    (a) The Kaibab Plateau
    (b) The Colorado Plateau
    (c) The Colorado River
    (d) Marble Canyon

    (Hint: See the first edition Slide Set.)

4. Mount Shasta is a composite volcano of the:
    (a) Coast Ranges
    (b) Sierra Nevada
    (c) Cascade Range
    (d) Basin and Range province

    (Hint: See Video 26.1.)

5. The Canyonlands of Utah were carved by:
    (a) The Colorado River
    (b) The Green River
    (c) The Colorado and Green rivers
    (d) The Snake River

    (Hint: See Video 27.1.)

6. Many ridges of the Appalachians were folded during the _____ orogeny, which occurred when Africa collided with Laurentia to form the supercontinent Pangaea:
    (a) Appalachian
    (b) Alleghenian
    (c) Piedmont
    (d) Blue Ridge

    (Hint: See Video 29.2.)

FINAL REVIEW

After reading this chapter, you should:

*   Know what topography, elevation, and relief are
*   Know the principal components of landscape
*   Know the major factors controlling landscape
*   Be familiar with the landscapes of North America
*   Understand how landscapes evolve

ANSWER KEY

LANDSCAPE TERMS AND CONCEPTS

1. Cuestas, hogbacks,   2. Badlands,   3. Tectonics,   4. Physical, chemical,
5. Negative feedback, 6. Elevation, 7. Erosion

PRACTICE MULTIPLE-CHOICE QUESTIONS

1. a, 2. c, 3. c, 4. c, 5. d, 6. a, 7. d, 8. a, 9. c, 10. a, 11. c, 12. a, 13. b, 14. b

17

The Oceans

The oceans, which cover almost three-quarters of Earth's surface, play an extremely important role in everything that occurs on our planet. Ninety-seven percent of all water on Earth is found in the oceans, and the hydrosphere is driven by what goes on there. This chapter examines many aspects of Earth's oceans: waves and tides, coasts and shorelines, the continental margins, the deep seafloor, and chemical and physical sedimentation in the sea.

◆ Coasts are the broad regions where land meets sea., and the shoreline is the line where the water surface intersects the shore. The major geological forces operating at the shoreline are waves and tides.

◆ Waves are created by the wind blowing over the surface of the water, transferring the energy of motion from air to water. Wave height increases as the wind speed increases, the wind blows longer, and the distance over which the wind blows the water increases. Waves travel as a form, but the water stays in the same place. Figure 17.3 on page 422 shows how waves travel as the water particles at the surface and beneath the waves move in circular vertical orbits.

A wave form can be described by its wavelength (the distance between crests), the wave height (the vertical distance between the crest and the trough), and the period, the time it takes for successive waves to pass a given point. The velocity of a wave equals the wavelength divided by the period.

Waves far from shore (called swell) are low, broad, and rounded. As swell approaches the shore, the waves break and form surf. The belt offshore where the breakers collapse is the surf zone. After breaking at the surf zone, the waves continue to move in, breaking again right at the shoreline. They run up onto the sloping front of the beach as swash, then back down again as backwash.

Far from shore, the lines of swell are parallel to one another but are usually at some angle to the shoreline. As the waves approach the beach, the rows of waves gradually bend to a direction more parallel to the shore; this bending is called wave refraction. Refraction results in more intense wave action (and hence more erosion) on projecting headlands and less intense action in indented bays, because the water becomes shallow more quickly around headlands than in the surrounding bays.

Although refraction makes waves more parallel to the shore, many waves still approach at a small angle. As the waves break on the shore, the swash moves up the beach slope perpendicular to this small angle. The backwash runs down the slope at a similar small angle but in the opposite direction. The combination of the two motions results in a trajectory that moves the water a short way down the beach. Sand grains carried by swash and backwash are moved along the beach in a zigzag motion known as longshore drift.

Waves approaching the shoreline at an angle can also cause a longshore current, a shallow-water current parallel to the shore. The water that moves with swash and backwash in and out from the shore at an angle creates a zigzag path of water particles that adds up to a net transport along the shore in the same direction as the longshore drift.

♦   Tides result from the gravitational pull of the Moon and the Sun on the water of the oceans. Figure 17.9 on page 426 shows how the Moon's gravity produces high and low tides. The Sun also causes tides. When the Moon, Earth, and Sun line up (at full Moon and new Moon), the gravitational pulls of the Sun and the Moon reinforce each other, producing the high tides called spring tides. The lowest tides, the neap tides, come in between, at first- and third-quarter Moon. Figure 17.10 on page 427 illustrates spring and neap tides.

Tides moving near shorelines cause currents that can reach speeds of a few kilometers per hour. As the tide rises, the water flows in toward the shore as a flood tide, moving into shallow coastal marshes and up small streams. As the tide starts to fall, the ebb tide moves out, and low-lying coastal areas are exposed again. Such tidal currents meander across and cut channels into tidal flats, the muddy or sandy areas that lie above low tide but are flooded at high tide.

♦   Waves, longshore currents, and tidal currents interact with the rocks and tectonics of the coast to shape shorelines into many forms. We can see these factors at work on beaches. A beach is a shoreline made up of sand and pebbles. The major parts of a beach are the offshore, bounded by the surf zone; the foreshore, which includes the surf zone and the tidal flat; the swash zone; and the backshore, which extends from the swash zone to the highest level of the beach.

The sand budget of a beach is determined by the inputs and outputs that occur through erosion and sedimentation. A beach gains sand from erosion of material along the backshore; from longshore drift and longshore current; and from rivers that enter the sea, bringing in sediment. The beach loses sand when winds carry sediments to backshore dunes, longshore drift and currents carry it downcurrent, and deep water transports it by currents and waves during storms. If the total input balances the total output, the beach is in equilibrium and keeps the same general form. If input and output are not balanced, the beach either grows or shrinks. Temporary imbalances are natural over weeks, months, or years, but over the long run, a beach tends to be in equilibrium.

◆   The topography of the shoreline is a product of tectonic forces elevating or depressing the Earth's crust, erosion wearing it down, and sedimentation filling in the low spots. The factors directly at work are uplift of the coastal region, which leads to erosional coastal forms; subsidence of the coastal region, which produces depositional coastal forms; the nature of the rocks or sediments at the shoreline; changes in sea level; the average and storm wave heights; and the heights of the tides, which affect both erosion and sedimentation

Erosion is active at tectonically uplifted rocky coasts. Cliffs and headlands jut into the sea, alternating with narrow inlets and irregular bays with small beaches. Waves crash against rocky shorelines, undercutting cliffs and causing huge blocks to fall into the water, where they are gradually worn away. As the sea cliffs retreat by erosion, isolated remnants called stacks are left standing in the sea, far from the shore. Erosion by waves also planes the rocky surface beneath the surf zone and creates a wave-cut terrace.

Sediment builds up in areas where tectonic subsidence depresses the crust along a coast. Such coasts are characterized by long, wide beaches and wide, low-lying coastal plains. Shoreline forms include sandbars, low-lying sandy islands, and extensive tidal flats. Long sandbars offshore may build up and become barrier islands, forming a barricade between open ocean waves and the main shoreline.

◆   Shorelines are sensitive to changes in sea level, which can change the approach of waves, alter tidal heights, and affect the path of longshore currents. Rise and fall of sea level can be local, a result of tectonic subsidence or uplift, or global, such as by continental glacial melting or growth. When sea level falls, areas that were offshore are exposed to agents of erosion. When sea level rises, river valleys are drowned, marine sediments build up along former land areas, and erosion is replaced by sedimentation. Drowned river valleys are one kind of estuary, which is a coastal body of water connected to the ocean and also supplied with fresh water from a river.

◆   Much of what we know about the deep-water portions of the oceans and the ocean floor has been learned in the past 30 years or so. To study these areas, geologists and oceanographers use deep-diving submersibles and instruments that allow them to sense the seafloor from a ship on the surface. These instruments include echo sounders, radar, and underwater cameras. Information is also gained from cores taken from holes drilled into the ocean floor.

◆   By comparing the topographic profiles of the Atlantic and Pacific oceans, we can learn much about the plate tectonic processes that produced the contrasting topographies of their ocean floors. Figure 17.24 on page 436 shows the topographic profile of the Atlantic Ocean. Figure 17.28 on page 440 shows the Pacific topographic profile.

◆   Continental margins comprise the shorelines, shelves, and slopes of the continents. There are two types: passive and active. Passive margins are found far from a plate boundary; they are called passive because there are no volcanoes and few earthquakes. At active margins, which are associated with subduction zones and transform faults, there is frequent volcanic activity and earthquakes

Continental shelves are broad and relatively flat at passive continental margins, narrow and uneven at active margins. Because continental shelves lie at shallow depths, they are exposed or submerged as a result of changes in sea level.

An interesting feature of the continental slope and rise is a turbidity current—a flow of turbid, muddy water down the slope. Turbidity currents can both erode and transport sediment. They start when occasional earthquakes trigger slumps of sediment draped over the edge of the continental shelf and onto the continental slope. The layers of sediment deposited by turbidity flows form a submarine fan, a fan-shaped deposit something like an alluvial fan on land.

Submarine canyons are deep valleys eroded into the continental shelf and slope. Although other types of currents have been proposed, turbidity currents are now believed to be responsible for the deeper parts of submarine canyons.

◆   The deep seafloor is constructed primarily by volcanism related to plate-tectonic motions and secondarily by sedimentation in the open sea. When plates grow by spreading from a mid-ocean ridge, huge quantities of basalt well up from the mantle, forming the oceanic lithosphere. As the plates spread away from the ridge, the lithosphere cools and contracts, lowering the seafloor. While this is happening, the basalt surface receives a steady rain of sediment from surface waters and gradually becomes mantled with deep-sea muds and other deposits.

Intense volcanic and tectonic activity occurs at mid-ocean ridges. The walls of the main rift valley are faulted and intruded with sills and dikes, and the valley floor is covered with flows of basalt and talus blocks from the valley walls, mixed with sediment settling from surface waters.

Hydrothermal springs form on the rift valley floor as seawater percolates into cracks and fractures in the basalt on the flanks of the ridge, is heated as it moves down to hotter basalt, and finally exits at the valley floor. Hydrothermal springs on the seafloor produce mounds of iron-rich clay minerals, iron and manganese oxides, and large deposits of iron-zinc-copper sulfides.

THE GEOLOGY OF THE OCEANS

The geology of the oceans is quite different from that of the continents. Whereas plate tectonics and erosion shape the continents, volcanism and sedimentation predominate in the ocean. Volcanism creates island groups in the middle of the oceans; arcs of volcanic islands near deep oceanic trenches; and mid-ocean ridges. Sedimentation shapes much of the rest of the ocean floor. Soft sediments of mud and calcium carbonate blanket the low hills and plains of the seafloor and accumulate on oceanic plates as they spread from mid-ocean ridges. As the plates move farther and farther from a ridge, they accumulate more and more sediment. Eventually the plates are swallowed up by subduction zones, which destroy the oceanic sediments by metamorphism and melting.

The oceans have no folded and faulted mountains like those on the continents. Instead, plate-tectonic deformation is restricted to the faulting and volcanism found at mid-ocean ridges and at subduction zones. Weathering and erosion are much less important in the oceans than on the land because there are no efficient fragmentation processes, such as freezing and thawing, or major erosive agents, such as streams. Deep-sea currents can erode and transport sediment, but they cannot effectively attack basaltic plateaus or hills.

The absence of tectonic deformation, weathering, and erosion over most of the seafloor means that far more details of the geologic record are preserved in layers of oceanic sediment than in continental sediments. The oldest parts of the oceanic record, however, are continuously erased by subduction. The oldest sediments preserved on today's ocean floor are Jurassic, only about 150 million years old. It takes about 150 million years for the crust created at mid-ocean ridges to spread across an ocean and be destroyed at a subduction zone.

The floor of the deep oceans away from mid-ocean ridges is a landscape of hills, plateaus, sediment-floored basins, and seamounts. Most of the thousands of volcanoes are submerged, but some rise to the surface. Seamounts and volcanic islands may be isolated, in clusters, or in chains. They may be formed along a mid-ocean ridge or where a plate overrides a mantle hot spot. Guyots are seamounts with flat tops, the result of erosion of an island volcano when it was above sea level.

Abyssal hills, plateaus, and low ridges are accumulations of volcanic rock. Many of these features form when the seafloor first opens at a rift valley. Others form as volcanic chains are created over hot spots. Although the deep seafloor is tectonically active at mid-ocean ridges, subduction zones, and hot spots, it is quiescent over much of the vast areas of abyssal hills and abyssal plains.

Coral reefs can be atolls, islands in the open ocean with circular lagoons enclosed within a more or less circular chain of islands, or they can form at continental margins. The outermost part of a reef is a slightly submerged, wave-resistant reef front composed of the interlaced skeletons of actively growing coral and calcareous algae, forming a tough, hard limestone. Behind the reef front, a flat platform extends into a shallow lagoon. An island may lie at the center of the lagoon. Figure 17.34 on page 449 shows how a coral reef evolves from a subsiding volcanic island.

◆   A blanket of sediment covers virtually the entire seafloor. The sediment is of two kinds: terrigenous muds and sands eroded from the continents, and biochemically precipitated shells of organisms that live in the sea. Terrigenous sedimentation on the continental shelf is produced by waves and tides. Biochemical sedimentation on the shelf results from the buildup of the calcium carbonate shells of clams, oysters, and many other organisms living in shallow waters.

Far from the continental margins, fine-grained terrigenous and biochemically precipitated particles suspended in seawater slowly settle from the surface to the bottom. These pelagic sediments are characterized by great distance from continental margins, fine particle size, and a slow settling mode of deposition. The most abundant biochemically precipitated particles are the shells of foraminifera, tiny single-celled animals that float in the surface waters of the sea. After the organisms die, their calcium carbonate shells fall to the bottom and accumulate as foraminiferal oozes. Another kind of biochemically precipitated sediment, silica ooze, is produced by sedimentation of the silica shells of diatoms and radiolaria. After burial on the seafloor, silica oozes are cemented into chert.

◆   This survey of the oceans shows that they are geologically complex, with distinctive structures, topographies, and sediments. Our studies of the ocean floor have given us an appreciation of the differences between continental geology and submarine geology. *(See the accompanying article on the geology of the oceans.)*

CHAPTER 17 OUTLINE

SAND BUDGET OF A BEACH

Match the lettered items to the appropriate category below.

INPUTS: _____   OUTPUTS: _____

a. Sediments transported downcurrent by longshore drift and current
b. Sediments eroded from backshore cliffs by waves
c. Sediments transported by rivers
d. Sediments transported to deep water by tidal currents and waves
e. Sediments transported to backshore dunes by offshore winds
f. Sediments eroded from upcurrent beach by longshore drift and current

CONTINENTAL MARGINS

In the space next to the numbered features, place an A if the feature is associated with an active continental margin, a P if it is associated with a passive continental margin.

1. _____ Far from a plate boundary
2. _____ Associated with a spreading ocean
3. _____ Narrow, uneven continental shelf
4. _____ Volcanoes absent
5. _____ Tectonically deformed
6. _____ West coast of South America
7. _____ East coast of North America
8. _____ Associated with a subduction zone
9. _____ Associated with a transform fault
10. _____ Continental shelf made up of flat-lying, shallow water sediments

PRACTICE MULTIPLE-CHOICE QUESTIONS

Circle the option that best answers the question.

1. Waves are generated by:
   (a) Solar energy
   (b) Wind blowing across the water surface
   (c) The tides
   (d) Heat from the underlying molten rock of the mid-ocean ridges

2. The wavelength of a wave in the open ocean is 300 meters and its height is 3 meters. The period of the wave is 10 seconds. At what velocity is the wave moving across the open ocean?
   (a) 10 meters per second
   (b) 23.1 meters per second
   (c) 30 meters per second
   (d) 100 meters per second

3. The area that includes the surf zone, the tidal flat, and the swash zone is the:
   (a) Foreshore
   (b) Offshore
   (c) Backshore
   (d) Tidal flat

4. The longshore drift of sand is caused by:
   (a) Rip currents
   (b) The backwash of the waves
   (c) The incoming rush of the waves
   (d) The angle of incidence of waves with the beach

5. The current that lies within the surf zone that moves parallel to shore is the:
   (a) Refraction current
   (b) Headland current
   (c) Longshore current
   (d) Rip current

6. The highest tides occur when:
   (a) The Sun, Moon, and Earth are all aligned
   (b) The Sun and Moon are at right angles to the Earth
   (c) The Earth is closest to the Sun
   (d) The Moon is in either its first or last quarter

7. Rapid coastal erosion always occurs where:
   (a) The coastal rocks are unconsolidated sands and clays
   (b) The coast is exposed to prevailing onshore winds and currents
   (c) Rapid subsidence of the shoreline takes place
   (d) Rapid uplift of the shoreline takes place

8. Sea stacks are:
   (a) Piles of sedimentary rocks near the shoreline
   (b) The erosional remnants of sea cliffs
   (c) Depositional features located near stream inlets along the shore
   (d) Erosional features associated with reefs

9. An example of an erosional coastal feature is a:
   (a) Wave-cut terrace
   (b) Sandy beach
   (c) Barrier island
   (d) Spit

10. Long sandbars that form offshore are called:
    (a) Wave-cut terraces
    (b) Sandy beaches
    (c) Barrier islands
    (d) Spits

11. An estuary is:
    (a) An area where large deposits of coarse material accumulate
    (b) Formed as a depositional feature consisting of fine sands and muds
    (c) An area eroded by wave action
    (d) A drowned river valley that contains both fresh and salt water

12. A major difference between an Atlantic type of ocean-floor profile and a Pacific one is that:
    (a) The Atlantic profile has a much narrower continental shelf
    (b) The Pacific profile has a much narrower continental shelf
    (c) Atlantic profiles tend to have more subduction zones
    (d) The deepest areas in the ocean are in the Atlantic Ocean

13. The continental shelf extends out from the shoreline to a depth of:
    (a) 50 meters
    (b) 100 meters
    (c) 200 meters
    (d) 2000 meters

14. The continental margin of the Pacific Northwest would be called:
    (a) Active
    (b) Passive
    (c) Inactive
    (d) Dormant

15. A turbidity current is:
    (a) An extension of the current of a river where it enters the ocean
    (b) A flow of muddy water down the continental slope and rise
    (c) A turbulent current found mainly on the deep ocean floor
    (d) A current created by an offshore trench

16. Abyssal hills, plateaus, and low ridges on the ocean floor consist of:
    (a) Basalt
    (b) Limestone
    (c) Terrigenous sediments
    (d) Pelagic sediments

17. Which of the following features is located farthest from the shoreline?
    (a) Mid-ocean ridge
    (b) Abyssal plain
    (c) Continental rise
    (d) Continental slope

18. The explanation of how a coral reef evolves from a subsiding volcanic island was first proposed by:
    (a) Henry David Thoreau
    (b) Charles Darwin
    (c) Bruce Heezen
    (d) Jacques-Yves Costeau

19. If we drilled down through an atoll, what rock(s) would we encounter?
    (a) Granite
    (b) Limestone
    (c) Predominantly metamorphic rocks
    (d) Basalt overlain by limestone

20. Foraminiferal oozes consist primarily of:
    (a) Feldspar grains
    (b) Silica
    (c) Limonite
    (d) Calcite

21. Radiolaria and diatoms consist of:
    (a) Feldspar grains
    (b) Silica
    (c) Limonite
    (d) Calcite

22. The oldest sediments preserved on the seafloor are:
    (a) 50 million years old
    (b) 100 million years old
    (c) 150 million years old
    (d) 200 million years old

CD RESOURCES

SUPPLEMENTAL DIAGRAMS

17.37 Properties of a wave
17.38 Various types of breakers in the surf zone
17.39 Wave energy in headlands and bays
17.40 Rip currents
17.41 Shoreline profiles as balance between tectonic forces and erosional/sedimentational processes
17.42 Tectonically active coasts versus coasts located away from plate boundaries

SUPPLEMENTAL PHOTOS

FIRST EDITION SLIDE SET

ANIMATIONS

EXERCISES

CD LAB

1. Name four kinds of breakers.

   (Hint: See the Chapter 17 supplemental diagrams.)

2. Shorelines profiles are a balance between tectonic uplift or subsidence and sedimentation or erosion. As tectonic subsidence of the seafloor increases, sedimentation:
   - (a) Increases
   - (b) Decreases

   (Hint: See Chapter 17 supplemental diagrams.)

3. Black sand beaches, such as are found in Hawaii, are composed of:
   - (a) The shells of calcereous organisms
   - (b) Black quartz
   - (c) Basalt
   - (d) Onyx

   (Hint: See Chapter 17 supplemental photos.)

4. The Great Barrier Reef, the largest modern reef in the world, is home to how many different species?
   - (a) 50
   - (b) 100
   - (c) 250
   - (d) 350

   (Hint: See the first edition Slide Set.)

5. Reef-building corals grow most rapidly in waters that are:
   - (a) Shallow, murky, and warm
   - (b) Shallow, clear, and warm
   - (c) Deep, clear, and cool
   - (d) Deep, clear, and warm

   (Hint: See the first edition Slide Set.)

FINAL REVIEW

After reading this chapter, you should:

- Understand the dynamics of wave motion and know what $V = L/T$ means
- Understand how tides and tidal currents occur
- Understand the structure and dynamics of a beach

- Understand the geologic processes that take place at shorelines
- Know some of the differences between the floors of the Atlantic and Pacific oceans
- Know the major features of continental margins
- Understand how deep ocean floors are formed
- Know about ocean sedimentation
- Know the key geologic differences between oceans and continents

Answer Key

SAND BUDGET OF A BEACH

Inputs: b, c, f   Outputs: a, d, e

CONTINENTAL MARGINS

1. P, 2. P, 3. A, 4. P, 5. A, 6. A, 7. P, 8. A, 9. A, 10. P

PRACTICE MULTIPLE-CHOICE QUESTIONS

1. b, 2. c, 3. c, 4. d, 5. c, 6. a, 7. d, 8. b, 9. a, 10. c, 11. d, 12. b, 13. c, 14. a, 15. b, 16. a, 17. a, 18. b, 19. d, 20. d, 21. b, 22. c

18 Earthquakes

An earthquake is one geological activity that gets everyone's attention. When the blocks of rock on either side of a fault slip suddenly, energy is released and the ground shakes. Earthquakes can cause extensive property damage as well as injuries and death.

This chapter examines how seismologists locate, measure, and try to predict earthquakes. It also discusses the effects of earthquakes and how they can be mitigated. We will see how the pattern of earthquakes helped lead to the development of plate-tectonic theory and how the locations of earthquakes are related to plate boundaries.

♦ The elastic rebound theory presents an explanation of why earthquakes occur. Figure 18.2 on page 000 shows the theory in schematic form. Two crustal blocks on either side of a transform fault are slowly forced to slide past each other. Friction along the fault prevents slip, and the crust is deformed. Strain builds up until the "frictional lock" is broken, a rupture occurs, and the blocks rebound to their previous, unstrained state. The rupture spreads, and an earthquake slip occurs over a section of the fault.

The elastic strain that slowly builds up over decades when two blocks are pushed in opposite directions is like the strain exerted on a rubber band when it is slowly stretched. The sudden release of strain in an earthquake, signaled by slip along a fault and the release of intense vibrations called seismic waves, is analogous to the violent rebound that occurs when the rubber band breaks. Elastic energy is being stored in the stretched rubber band, and it is this energy that is suddenly released in the backlash. In the same way, elastic energy accumulates and is stored over many decades in rocks under strain. The energy is released at the moment the fault ruptures and is radiated as seismic waves in the few minutes of an earthquake. The point at which the slip initiates is called the focus and the geographic point on Earth's surface directly above the focus is the epicenter.

♦   A seismograph is an instrument used to record the seismic waves generated by earthquakes. A seismograph installed anywhere on Earth will, within a few hours, record the passage of seismic waves generated by earthquakes somewhere. The waves travel from the earthquake focus through the Earth and arrive at the seismograph in three distinct groups: primary waves or P waves, secondary waves or S waves, and finally surface waves. Both P and S waves travel through Earth's interior; surface waves travel around Earth's surface.

P waves in rock are analogous to sound waves in air. Both P waves and sound waves are compressional waves, which travel through solids, liquids, or gases as a succession of compressions and expansions. P waves can be thought of as push-pull waves: they push or pull particles of matter in the direction of their path of travel.

S waves travel through solid rock at about half the speed of P waves. They are shear waves that push material at right angles to their path of travel. Shear waves do not exist in liquids or gases.

Surface waves are confined to the Earth's surface and outer layers because, like waves on the ocean, their existence depends on a free surface to ripple. Their speed is slightly less than that of S waves. One type of surface wave sets up a rolling motion in the ground; another type shakes the ground sideways.

♦   Seismic waves enable geologists to locate earthquakes and determine the nature of faulting. The principle used to locate an earthquake's epicenter is similar to the principle used to deduce the distance to a lightning bolt based no the time interval between the flash of light and the sound of thunder—the greater the distance to the bolt, the larger the time interval. The lightning flash may be likened to the P waves and the thunder to the slower S waves.

To determine the approximate distance from a seismograph to an earthquake's epicenter, seismologists read from a seismogram the amount of time that elapsed between the arrival of the first P waves and the later arrival of the S waves. Then they use a table or a graph to determine the distance from the seismograph to the epicenter. If they know the distances from three or more stations, they can pinpoint the epicenter. They can also deduce the time of the shock at the epicenter because the arrival time of the P waves at each station is known, and from a graph or table it is possible to determine how long the waves took to reach the station.

♦   In addition to being able to locate earthquakes, the seismologist must also determine their magnitude. An earthquake's magnitude is a primary factor in its destructiveness. The sizes and numbers of earthquakes in a given area also indicate the degree of tectonic activity.

Two primary measures of magnitude are commonly used today. The Richter magnitude depends on the amplitude (size) of the ground movement caused by seismic waves. Two earthquakes that differ in size of ground motion by a factor of 10 differ in magnitude by 1 Richter unit. The ground motion of an earthquake of magnitude 3 is 10 times that of an earthquake of magnitude 2. The moment magnitude reflects what happened at the earthquake source rather than how much the ground shakes at a distant point. It depends on the product of the slip of the fault when it broke, the area of the fault break, and the rigidity or stiffness of the rock. Although both the Richter method and the moment method produce roughly the same numerical values, the moment magnitude is more closely related to the cause of an earthquake than to its effect and is therefore more useful to scientists.

Earthquake magnitude does not necessarily describe the destructiveness of a particular earthquake because a magnitude 8 earthquake 2000 km from the nearest city might cause no human or economic damage, while a magnitude 6 quake immediately beneath a city could cause serious damage. The Modified Mercalli Intensity Scale assigns a measure of the destructiveness of an earthquake rather than its magnitude or physical size. It is a qualitative scale based on the observed effect on people and damage to structures.

◆   We also want to know the orientation of the fault plane and the direction of slip in an earthquake and how those characteristics fit into the regional pattern of crustal forces. In some directions from an earthquake, the initial ground movement is a push away from the focus. In other directions, the initial ground movement is a pull toward the focus. These differences reflect the fact that the slip on a fault looks like a push if you view it from one direction but like a pull if you view it from another. From these "pushes" and "pulls," the fault orientation and the slip can be inferred, and geologists can deduce whether the crustal forces that triggered the earthquake were compressional, tensional, or shear.

◆   Seismologists have now been able to discern patterns in the distribution of earthquakes of various types. They have explained these patterns within the framework of plate-tectonics theory—and thereby provided tremendous support for the theory. *(See the accompanying article on earthquakes and plate tectonics.)*

◆   Each year, 800,000 little tremors that are not felt by humans are recorded. About 100 earthquakes occur each year with Richter magnitudes between 6 and 7. A great earthquake, one with a Richter magnitude exceeding 8, occurs somewhere in the world about once every 5 to 10 years.

It is fortunate that most earthquakes are small. Damage to buildings and other structures near the epicenter begins at magnitude 5 and increases to nearly total destruction when the magnitude is greater than 8. Two California earthquakes—the 1989 Loma Prieta earthquake (magnitude 7.1) and the 1994 Northridge earthquake (magnitude 6.9)—were among the costliest disasters in the history of the United States. As damaging as these earthquake were, they released less than a hundredth the energy of some truly great earthquakes, such as those in San Francisco (1906, 8.3), Tokyo (1923, 8.2), Chile (1960, 8.6), Alaska (1964, 8.6), and China (1976, 7.8).

♦   Earthquakes cause destruction in several ways: the collapse of buildings; fires ignited by ruptured gas lines or downed electrical power lines; avalanches; and tsunamis, towering waves that travel across the ocean at speeds of up to 800 km per hour and form 20-m-high walls of water when they r each the coast..

♦   Can anything be done to reduce earthquake hazards? A seismic-risk map provides a basis for organizing local earthquake protection programs consonant with the degree of danger. In high-risk areas, building codes should require earthquake-resistant structures. Construction on unstable soils or in avalanche-prone areas should also be regulated. Box 18.2 on page 000 presents steps you can take to protect yourself and your family in an earthquake.

♦   A few earthquakes are claimed to have been predicted in the China, the United States, and the former Soviet Union, but the vast majority go unpredicted. Scientists in the United States, Japan, China, Russia, and other countries are engaged in an intensive search for premonitory indicators to help predict the time and place of a destructive earthquake. Possible indicators being examined include an unusual increase in the frequency of smaller earthquakes in a region before a main shock; a rapid tilting of the ground or other form of surface deformation; an unusual aseismic slip (smooth, slow sliding along a fault rather than the sudden slip of an earthquake) on a fault; an episode of stretching of the crust across a fault; changes in the physical properties of rock in the vicinity of a fault, such as its ability to conduct an electric current or its P-wave velocity; and changes in the level of water in wells. All these phenomena have been observed in various combinations, but not with the consistency and reliability that a useful prediction method requires.

The seismic gap method of earthquake prediction is based on the idea that earthquakes result from the accumulation of strain caused by the steady motion of the plates along faults. When some critical level of strain is reached, the brittle lithosphere breaks. The cycle of slow accumulation of strain and its sudden release in an earthquake recurs again and again, and the average interval varies from place to place. According to the seismic gap method, the most likely place for an earthquake to occur is at a locked portion of a fault where the time since the last earthquake has reached or exceeds the average interval between earthquakes in that location.

If earthquakes could be predicted so that a warning could be issued hours or days before the shock, casualties could be reduced significantly. Unfortunately, a reliable method of predicting earthquakes within a few days and with very few false alarms has yet to be found. All that seismologists can do today is provide estimates of probability.

EARTHQUAKES AND PLATE TECTONICS

Seismologists have known for decades that earthquakes tend to occur in "belts" (see Figures 18.14 and 18.15 on page 470). One of the best known belts corresponds to the "Ring of Fire" surrounding the Pacific Ocean. Because we can detect the more numerous small earthquakes and have improved methods of locating epicenters, seismic belts can now be defined so accurately that they can be correlated with geologic features.

Almost all divergent plate boundaries occur on the seafloor. The narrow belts of mid-ocean earthquakes coincide with mid-ocean ridge crests. When the topography of mid-ocean ridges is examined in detail, the ridges are often found to be segmented, the segments being offset by transform faults. Earthquake epicenters also line the transform faults between the offset ridge segments. Moreover, the fault mechanisms of the ridge-crest earthquakes, revealed by analysis of the first P-wave motion, are normal, with the faults striking parallel to the trend of the ridges. Normal faulting indicates that tensional (or pull-apart) forces are at work. This is why rift valleys develop in the ridge crests. Seismologists, using the independent evidence of their seismic records, found that mid-ocean ridges define plate boundaries where plates are being pulled apart. They also found that the earthquakes that coincide with transform faults show strike-slip mechanisms—just as one would expect where plates slide past each other. Seismology gives elegant support to the idea that plates spread apart at mid-ocean ridge crests.

The earthquakes that originate at depths greater than about 100 km are found to coincide with continental margins or island chains that are adjacent to ocean trenches and young volcanic mountains. These features define a boundary where plates collide. Almost all deep-focus earthquakes occur along the inclined plane of a subducted plate, where it plunges back into the mantle beneath an overriding plate. An analysis of the initial P waves of these earthquakes reveals that the earthquakes in the area are produced by thrust faulting, signifying that compressive forces are at work, as would be the case at a collision boundary.

Although most earthquakes occur at plate boundaries, a small percentage originate within plates. Their foci are relatively shallow, and the majority occur on continents. Apparently, strong crustal forces can still occur and cause faulting within the lithospheric plates, far from modern plate boundaries.

♦ A puzzling new pattern of earthquakes has arisen in southern California. Since 1971, the area has had a barrage of moderate earthquakes on a system of thrust faults beneath Los Angeles. Many of these "blind thrust" faults are not visible on the surface and are not otherwise known until an earthquake occurs. How these recently activated thrust faults relate to plate movements and to the nearby San Andreas strike-slip fault is a question geologists are investigating.

Another source of concern is evidence that many great earthquakes have occurred in northern California, Oregon, Washington, and British Columbia over the past few thousand years. Although the severe earthquake potential of this region was recognized only recently, it should not come as a surprise. The Pacific Northwest is a convergent boundary, where the Juan de Fuca Plate is being subducted under the North American Plate, not as a continuous motion but as an accumulation of repeated slips with accompanying great earthquakes. Geologists now recognize that this densely populated region is at risk for another great earthquake.

CHAPTER 18 OUTLINE

SEISMIC WAVES

Decide which type(s) of seismic waves is (are) being described or depicted in the lettered options. Place the letter of each option after one or more of the wave types listed.

P waves _____   S waves _____   Surface waves _____

A. "Push-pull" waves
B. Push materials at right angles to their path of travel
C. Travel through Earth's interior
D. Slowest of the three wave types
E. Analogous to sound waves in air
F. Travel through solid rock at about 5 km per second
G. Need to ripple to exist
H. Compressional waves
I. Analogous to ocean waves
J. Shear waves

PRACTICE MULTIPLE-CHOICE QUESTIONS

Circle the option that best answers the question.

1. Large earthquakes are caused predominantly by:
   (a) Volcanism and the intrusion of magma
   (b) Major landslides and avalanches
   (c) Horizontal and vertical movements of the Earth's crust
   (d) The release of strains produced by tidal forces

2. The movement of adjacent blocks past one another along a fault plane is described by the:
   (a) Elastic rebound theory
   (b) Seismic gap method
   (c) Richter magnitude
   (d) Modified Mercalli Scale reading

3. The point in the subsurface at which slip is initiated along a fault plane is:
   (a) The epicenter of the earthquake
   (b) The focus of the earthquake
   (c) Found by calculating the Richter magnitude
   (d) Always more than 100 km deep

4. The instrument used to measure earthquake waves is a:
   (a) Seismograph
   (b) Seismogram
   (c) Global Positioning System (GPS) station
   (d) Richter scale

5. The order in which seismic waves arrive at a fixed point is (from first to last recorded):
   (a) Surface wave, S wave, P wave
   (b) S wave, P wave, surface wave
   (c) S wave, surface wave, P wave
   (d) P wave, S wave, surface wave

6. Which of the following is not a characteristic of S waves:
   (a) They arrive before surface waves
   (b) They travel slower than P waves
   (c) They are not propagated through liquids
   (d) They are the slowest type of seismic waves

7. A subjective scale used to measure earthquake effects is the:
   (a) Richter scale
   (b) Modified Mercalli scale
   (c) Moment Magnitude scale
   (d) Logarithmic scale

8. Earthquake A has a magnitude of 5.5 and earthquake B has a magnitude of 7.5. Which of the following statements is true?
   (a) The ground moved exactly twice as much in earthquake B as it did in A
   (b) The ground moved 13 times (5.5 + 7.5) as much in earthquake B as it did in A
   (c) The ground moved 100 times as much in earthquake B as it did in A
   (d) The relative amount of ground movement cannot be determined due to insufficient information

9. Deep-focus earthquakes tend to occur:
   (a) Within plates
   (b) At convergent plate boundaries
   (c) At divergent plate boundaries
   (d) At transform fault boundaries

10. The country that has done the most to prepare for earthquakes is:
    (a) The United States
    (b) China
    (c) Russia
    (d) Japan

11. The seismic gap method of earthquake prediction is based on the observation that:
    (a) Earthquakes of a given magnitude recur at regular intervals
    (b) The probability of a major earthquake decreases with time since the last earthquake of similar magnitude
    (c) The probability of a major earthquake increases with time since the last earthquake of similar magnitude
    (d) Major earthquakes are most likely to occur in the deep mantle

12. Outside of California, the area of the United States most likely to experience a major earthquake is:
    (a) The Pacific Northwest
    (b) Florida
    (c) New England
    (d) The Great Lakes region

CD RESOURCES

SUPPLEMENTAL DIAGRAMS

18.25 Principle of the pendulum-mounted seismograph
18.26 Principle of a spring seismograph
18.27 Spring-mounted seismograph to record vertical ground motion
18.28 Principle of a modern seismograph
18.29 Energy released in earthquakes compared with other phenomena
18.30 Epicenters of large-magnitude earthquakes worldwide, 1897–1992
18.31 Isoseismal contours for the 1906 San Francisco, 1971 San Fernando, and 1811–1812 New Madrid earthquakes
18.32 Seismic gaps for the Circumpacific region in 1989
18.33 Odds of major earthquakes on San Andreas fault in next 30 years
18.34 Seismic-risk map of the United States
18.35 Hazard map of the United States
18.36 Cascadia subduction zone

SUPPLEMENTAL PHOTOS

18.37 Aerial view, offset streams across San Andreas fault
18.38 Aerial view, San Andreas fault, Carrizo Plain, California
18.39 Seismographs

18.40 Earthquake epicenters, 1972–1989, San Francisco Bay Area

18.41 View of damage to city in 1906 San Francisco earthquake

18.42 Tsunami damage along waterfront at Kodiak; March 1964 earthquake

18.43 Surge wave damage to trees, Port Valdez; March 1964 earthquake

18.44 Landslide damage, Turnagain Heights, Anchorage; March 1964 earthquake

18.45 Graben in downtown Anchorage; March 1964 earthquake

18.46 Sherman Glacier, Alaska, before March 1964, earthquake

18.47 Sherman Glacier, Alaska, after March 1964, earthquake

18.48 Damage to twin bridges; 1989 Loma Prieta earthquake

18.49 Collapsed section of I-880 Cyprus viaduct; 1989 Loma Prieta earthquake

18.50 Bridge buckled during January 1994 Northridge earthquake

18.51 House damaged during January 1994 Northridge earthquake

18.52 House demolished during January 1994 Northridge earthquake

18.53 House collapsed during January 1994 Northridge earthquake

18.54 Housing collapsed during January 1994 Northridge earthquake

18.55 House collapsed during January 1994 Northridge earthquake

18.56 Highway overpass collapsed during January 1994 Northridge earthquake

18.57 Store destroyed during January 1994 Northridge earthquake

18.58 Highway overpass collapsed during January 1994 Northridge earthquake

18.59 Overturned auto on highway damaged during January 1994 Northridge earthquake

18.60 Landslide, Kobe; Great Hanshin Earthquake of 1995

18.61 Landslide, Kobe; Great Hanshin Earthquake of 1995

18.62 Landslide, Kobe; Great Hanshin Earthquake

18.63 Slides caused by heavy rains required evacuation of residents; Great Hanshin Earthquake of 1995

18.64 Slides caused by heavy rains required evacuation of residents; Great Hanshin Earthquake of 1995

18.65 Kobe resident filling containers with water after Great Hanshin Earthquake

18.66 Collapsed section of Hanshin Expressway, Kobe; Great Hanshin Earthquake

18.67 Destruction of subway by sinking of Daikai Street; Great Hanshin Earthquake

18.68 Damage at Hyogo Port, Kobe; Great Hanshin Earthquake

18.69 Widespread fires from broken gas mains; Great Hanshin Earthquake

18.70 Mitsubishi Bank, Kobe; Great Hanshin Earthquake

18.71 Disruption of hydrofoil service; Great Hanshin Earthquake

18.72 Port Island, Kobe; Great Hanshin Earthquake

18.73 Soil liquefaction at elementary school playground Great Hanshin Earthquake

18.74 Collapse of Rokko Michi train station, Kobe Great Hanshin Earthquake

18.75 Rokko Michi train station restored to full operation Great Hanshin Earthquake

FIRST EDITION SLIDE SET

Slide 18.1 Change in stress induced by earthquakes

SECOND EDITION SLIDE SET

Slide 18.1 Apartment building collapsed on a row of parked cars in Northridge, California, after the January 1994, magnitude 6.7 earthquake

Slide 18.2 Inside the bookstore at California State University at Northridge after the January 1994, magnitude 6.7 earthquake

Slide 18.3 Chimneys toppled by the January 1994, magnitude 6.7 earthquake in Northridge, California

Slide 18.4 Earthquake-resistant wood-framed chimney design on home in Northridge, California, after the January 1994, magnitude 6.7 earthquake

Slide 18.5 Cement blocks in a sidewalk at Central Avenue and Powell Street in Hollister, California, are sheared by creep on the Calaveras fault

ANIMATIONS

Earthquake: Normal Fault
Earthquake: Reverse Fault
Earthquake: Strike-Slip Fault

ILLUSTRATED GLOSSARY

EXERCISES

Earthquakes and Plate Boundaries
Earthquakes at Convergent Plate Boundaries

CD LAB

1. Which segment of the San Andreas fault has the highest probability of a major earthquake in the next 30 years?
   (a) Choachella Valley
   (b) Mojave
   (c) San Francisco Peninsula
   (d) Parkfield

   (Hint: See the Chapter 18 supplemental diagrams.)

2. Which earthquake had the greatest magnitude?
   (a) 1960 Chile
   (b) 1906 San Francisco
   (c) 1964 Alaska
   (d) 1989 Loma Prieta

   (Hint: See the Chapter 18 supplemental diagrams.)

3. Damage in the Great Hanshin Earthquake of 1995 included:
    (a) Landslides and fires
    (b) Liquefaction
    (c) Damage to ports and subways
    (d) All of the above

(Hint: See the Chapter 18 supplemental photos.)

4. How far away from the epicenter did the 1992 Landers earthquake trigger earthquakes and activity in hot springs and geysers?
(Hint: See the first edition Slide Set.)

FINAL REVIEW

After reading this chapter, you should:

- Know the definition of an earthquake
- Know some of the methods and tools used to study earthquakes
- Know the three types of seismic waves
- Understand the relationship between earthquakes and plate tectonics
- Understand the ways in which earthquakes cause destruction
- Understand how we can predict and limit the damage of earthquakes

ANSWER KEY

SEISMIC WAVES

P waves: A, C, E, F, H, K
S waves: B, C, J, L
Surface waves: D, G, I, M

PRACTICE MULTIPLE-CHOICE QUESTIONS

1. c, 2. a, 3. b, 4. a, 5. d, 6. d, 7. b, 8. c, 9. b, 10. d, 11. c, 12. a

19

Exploring Earth's Interior

Geologists who study the deep interior of the Earth have no tangible material to analyze and must infer the properties and behavior of matter deep inside the Earth from phenomena observable at the surface. Geologists study Earth's interior by measuring the speed of seismic waves, the flow of heat from great depths to the surface, and the properties of Earth's magnetic field. In this chapter, we will see how geologists use the behavior of seismic waves to infer the kinds of materials that make up Earth's interior. We will also see how heat flow and magnetism provide clues to the structure and composition of the interior.

◆   The velocity at which any wave travels depends on the material it passes through. Light waves travel more quickly through air than through glass, and seismic P and S waves travel more rapidly through basalt than through granite. Geologists calculate the velocity of a P or S wave by measuring the time of travel and dividing the distance traveled by the travel time. Any differences detected for different paths can be used to infer the properties of materials the waves have passed through. The first experiments measuring the travel time of P waves from earthquakes found two P waves, one traveling faster than the other. This result led to the discovery of the Earth's crust: The slow wave could be attributed to a path through the crust, the fast wave to a path through the mantle below.

When waves encounter more than one material, some of the waves are reflected at the boundary between two materials and others are transmitted into the second material. The waves that cross the boundary between two materials refract, because their velocity is not the same in the second material as in the first. Seismologists study how seismic waves are refracted and reflected, and this behavior reveals certain features of Earth's interior.

P and S waves do not follow a straight path through the interior from the focus of an earthquake to a distant seismograph. This tells us that the Earth is layered and is made up of many materials that conduct P and S waves at different velocities. The waves bend as they go from layer to layer, and so their paths through the interior are curved.

Geologists have determined the paths of P and S waves and their corresponding travel times by studying the seismographic records of earthquakes all over the world. When an earthquake occurs, it produces P waves that radiate in all directions from the focus. One of those waves will encounter the boundary between the mantle and the outer core and will be reflected back to the surface. Slightly to either side of that wave is a P wave that will hit the same boundary at a slightly different angle and be refracted down into the outer core. This wave will continue traveling until it encounters the boundary between the outer core and the inner core and is reflected back through the mantle–outer core boundary to the surface. This wave emerges at the surface at a greater distance from the earthquake than the first wave. The area between the two waves is called the P-wave shadow zone because no P waves reach the surface there. (Figure 19.2a shows how this works in detail.) S waves behave in basically the same way, but the wave that is refracted into the outer core is absorbed and never reaches the surface, because S waves do not travel through liquids (see Figure 19.2b).

◆   Seismological observations and laboratory measurements of the properties of materials paint a picture of Earth's interior: a zoned planet whose major components are a metallic iron core and a rocky mantle. A thin, brittle, lightweight crust—the end product of the differentiation process—caps the mantle.

The crust, Earth's outermost layer, ranges in thickness from about 5 km under the oceans to as much as 65 km under the continents. Correlations of wave velocity and rock type show that the crust consists mostly of granitic rocks, except on the floor of the deep ocean, where it consists entirely of basalt and gabbro. Below the crust is a sharp boundary between the crust and the mantle called the Mohorovicic discontinuity (Moho for short). The indications that the crust is less dense than the underlying mantle are consistent with the theory that the crust is made up of lighter materials that floated up from the mantle.

The principle of isostasy states that continents are less dense than the mantle and float on it. How can continents float on solid rock? Although the mantle is solid, it appears that over long periods, it has little strength and behaves like a viscous liquid when it is forced to support the weight of continents and mountains. As a large mountain range forms, it slowly sinks under gravity and the crust bends downward. When enough of a root bulges into the mantle, the mountain floats. If a mountain range is reduced by erosion, the weight on the crust is lightened, less root is needed for flotation, and the root floats up until both the root and the mountain disappear. This process is called isostatic rebound.

Below the crust is the mantle, which is made up of several zones (see Figure 19.7). The topmost is the solid lithosphere, a slab up to 100 km thick in which the continents are embedded. Below the lithosphere is the partially molten asthenosphere, which ends at a depth of about 200 km. At depths of about 400 km and 670 km, there are two transition zones in which atoms are forced into closer packing. The lower mantle extends from a depth of 700 km to the core.

Earth's core is made up of a liquid outer core (2900 to 5100 km) and a solid inner core (5100 to 6370 km). Seismic wave research and laboratory experiments have led geologists to conclude that Earth's core is composed mostly of iron, molten in the outer core and solid in the inner core.

♦   The interior of the Earth is extremely hot, the exterior relatively cool. The Earth's temperature is affected by two processes, conduction and convection. Conduction mechanically transfers vibrational motion from material at high temperatures to material at low temperatures (see Figure 19.9). Convection occurs when a substance is heated unevenly. The hot, less dense portion rises through the cooler, denser portion, while at the same time the cooler, denser portion sinks.

♦   One of the most important remaining questions about plate tectonics is the mechanism that moves the plates over Earth's surface. Although scientists disagree about the exact mechanism, they do agree that convection plays a central role. *(See the accompanying article on convection and plate tectonics.)*

CONVECTION AND PLATE TECTONICS

The British geologist Arthur Holmes was among the first to propose convection as the driving mechanism of continental drift. When he advanced his theory in the 1930s, he was 30 years ahead of his time. His idea had to wait for confirmation until the seafloor was extensively explored after World War II, which led to the concept of seafloor spreading in 1963.

Seafloor spreading and plate tectonics are direct evidence of convection at work. The rising hot matter under mid-ocean ridges builds new lithosphere, which cools as it spreads away from the ridge. Eventually, it sinks back into the mantle, where it is reabsorbed (see Figure 19.10).

Geologists disagree over exactly how convection drives the motion of plates. Some believe that only the upper few hundred kilometers of the mantle are subject to convection, which implies that the upper and lower mantles do not mix. Others think that the whole mantle is involved, still others that rising narrow mantle plumes beneath hot spots provide the driving force for convection.

Regardless of the specifics, geologists now believe that the movement of heat from the interior to the surface as the seafloor spreads is an important mechanism by which Earth has cooled over geologic time. They also recognize that the subduction of oceanic crust into the mantle is a mechanism that recycles materials that have resided at the surface for a few hundred million years.

♦   Geologists have attempted to infer the temperature of Earth's interior at various depths by combining the temperature of lava that emerges in volcanoes, laboratory data on the temperatures at which rocks melt, and information from seismology. A curve of temperature versus depth is called a geotherm; the text presents one possible geotherm in Figure 19.11.

♦   Geologists use Earth's magnetic field as a tool to examine Earth's interior. Earth's magnetic field behaves as if a small but powerful permanent bar magnet were located near the center of the Earth and inclined about 11° from the Earth's axis of rotation. There is one problem with this model, however: Heat destroys magnetism, and materials lose their permanent magnetism at temperatures above about 500°C. Material below depths of about 20 or 30 km in the Earth, therefore, cannot be magnetized because the temperatures are too high.

Another way to create a magnetic field is with electric currents. Dynamos in power plants make electricity by means of an electrical conductor in the form of a coil of copper wire rotated through a magnetic field. The rotation is driven by steam or falling water.

Geologists now think that such a dynamo may exist in Earth's liquid iron outer core. Scientists speculate that the liquid iron is stirred into convective motion by heat generated from radioactivity in the core. By a process not completely understood, this motion is thought to produce both the electric currents and the magnetic field needed to sustain a dynamo in the core, from which a magnetic field emanates to the surface.

♦   In the early 1960s, it was discovered that Earth's magnetic field has reversed polarity many times in the past. Many very hot, magnetizable materials become magnetized in the direction of the surrounding magnetic field as they cool below about 500°C. Groups of atoms of the material align themselves in the direction of the magnetic field when the material is hot. Once the material cools, these atoms are locked in place and therefore are always magnetized in the same direction. This is called thermoremanent magnetization, and this property of igneous rocks can be used to determine the direction of Earth's magnetic field at the time a rock was magnetized as it cooled.

Some sedimentary rocks take on a depositional remanent magnetization. Marine sedimentary rocks are formed when particles of sediment that have settled through the ocean to the seafloor become lithified. Magnetic grains among the particles would have become aligned in the direction of Earth's magnetic field as they fell through the water, and this orientation would be incorporated into the rock when the particles became lithified.

♦   Erratically, but roughly every half-million years, Earth's magnetic field changes polarity, taking perhaps a few thousand years to die down and then build up in the opposite direction. Reversals are clearly indicated in the fossil magnetic record of layered lava flows. The direction of remanent magnetism can be obtained for each layer, and in this way the time sequence of reversals of the field—that is, the magnetic stratigraphy—can be deduced. This information is of use to archeologists and anthropologists as well as geologists.

CHAPTER 19 OUTLINE

COMPOSITION AND STRUCTURE OF EARTH'S INTERIOR

The figure on the next page represents a seismologic interpretation of various information about the travel of P waves and S waves. To seismologists, the changes that occur in P-wave and S-wave velocities at different depths in the Earth reveal the sequence of layers that make up Earth's interior. Answer the following questions about this seismologic interpretation of seismic wave data. **Study Hint:** *This material is covered on pages 488–495 of the textbook.*

THE CRUST

1. The crust is an average of about 35 km thick under the _____.

2. By compiling a library of information about the compositional materials of the Earth and the speed at which seismic waves pass through them, seismologists have surmised that the continental crust consists mostly of _____ rocks and that the crust of the ocean floor consists of _____ and _____.

3. P-wave velocity increases abruptly below the crust, suggesting a sharp boundary, which has been named the _____. The increase in velocity also suggests that the rock in the crust is *(less dense, more dense)* than rock in the mantle.

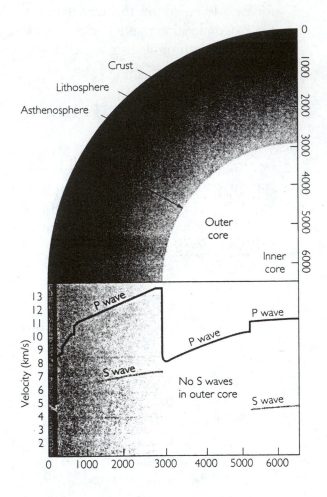

THE MANTLE

1. Seismologists have deduced that the lithosphere is solid because S waves *(do, do not)* pass through it and *(are, are not)* absorbed.

2. The theory of plate tectonics and the idea that the asthenosphere is the probable source of much basaltic magma are bolstered by the belief that the asthenosphere contains _____.

3. At the greatest depths, the increased velocity of S waves cannot be explained by changes in chemical composition, but rather by a repacking of _____ that could be caused by the _____ at that depth.

THE CORE

1. It is inferred that the core is solid at a depth of 5100 km because P-wave velocities suddenly _____ at that point.
2. Using various geological and astronomical data and experiments, scientists have inferred that the core consists entirely of _____.

PRACTICE MULTIPLE-CHOICE QUESTIONS

Circle the option that best answers the question.

1. Which of the following statements is *not* true?
    (a) Light waves and sound waves change velocity when they pass through different materials
    (b) Seismic waves change velocity when they pass through different materials
    (c) P and S waves can be reflected or refracted
    (d) Light and sound waves can be refracted, but seismic waves cannot

2. Information about Earth's interior composition and structure is obtained from:
    (a) Seismic waves produced by earthquakes
    (b) The analysis of inclusions brought up from the core
    (c) Deep-sea sediments
    (d) Magnetized basalts

3. The S-wave shadow zone is created by:
    (a) An increase in the velocity of S waves as they cross from mantle to core
    (b) The failure of S waves to be propagated through the outer core
    (c) The failure of S waves to be propagated through the inner core
    (d) An abrupt P-wave velocity decrease at the crust–mantle boundary

4. The principle of isostasy states that:
    (a) Continents float on the liquid mantle
    (b) Continents are less dense than the mantle and float on it
    (c) Continents are denser than the mantle and float on it
    (d) Continents float on the liquid outer core

5. When a mountain range is reduced by erosion, the weight on the crust is lightened, less root is needed for flotation, and the root floats up until both the root and the mountain disappear. This process is called:
    (a) Convection
    (b) Conduction
    (c) Isostatic rebound
    (d) The principle of isostasy

6. Within the asthenosphere, the velocity of S waves:
    (a) Equals that of P waves
    (b) Increases slightly
    (c) Increases by almost two times
    (d) Decreases

7. The Moho is the boundary between the:
    (a) P-wave and S-wave shadow zones
    (b) Crust and mantle
    (c) Lithosphere and asthenosphere
    (d) Core and mantle

8. The most significant occurrence at the mantle-core boundary is:
    (a) All P waves are reflected back to the surface
    (b) All S waves are reflected back to the surface
    (c) S waves are no longer transmitted into the core
    (d) P waves are no longer directly transmitted into the core

9. As you move from the outer edge of the Earth toward its center, the order in which the layers are encountered is:
    (a) Crust, mantle, outer core, inner core
    (b) Mantle, outer core, inner core, crust
    (c) Crust, mantle, inner core, outer core
    (d) Outer core, inner core, crust, mantle

10. The Earth's inner core is thought to be composed of:
    (a) Solid iron
    (b) Molten ultramafics
    (c) High-density carbonates
    (d) Various types of silicates not observed at the surface

11. The cooling of the Earth occurs most rapidly by:
    (a) Conduction
    (b) Convection
    (c) Radiative heat transfer
    (d) Cold seawater removing heat at mid-ocean ridges

12. The temperature at which iron loses its magnetic field is about:
    (a) 100°C
    (b) 250°C
    (c) 500°C
    (d) 4000°C

13. Magnetic reversals are best preserved in:
    (a) Granitic magmas located near the magnetic equator
    (b) Wind-blown sandstones formed in temperate climates
    (c) Basalts formed at mid-ocean ridges
    (d) Limestones deposited in shallow, fresh-water lakes

CD RESOURCES

SUPPLEMENTAL DIAGRAMS

19.17 Reflection and refraction of P and S waves
19.18 Cross section of Earth's gross structure revealed by earthquake studies
19.19 Crustal thicknesses in North America
19.20 Behavior of materials as solids and as viscous fluids
19.21 Response of a compass needle to slow changes in Earth's magnetic field

FIRST EDITION SLIDE SET

Slide 19.1 Refraction of light
Slide 19.2 Hotter and colder regions of Earth's mantle and crust

SECOND EDITION SLIDE SET

Slide 19.1 Pallasite (stony-iron) meteorite
Slide 19.2 Iron filings align along magnetic lines of force from a bar magnet
Slide 19.3 Seismic tomography shows evidence of the subducting Farallon Plate
Slide 19.4 Geoid of the United States
Slide 19.5 Geoid of the world

ILLUSTRATED GLOSSARY

EXERCISES

Exploring Earth's Interior

CD LAB

1. Crustal thicknesses in North America are generally:
    (a) Greater under mountain ranges
    (b) Greater near the coasts
    (c) About the same over the entire continent

   (Hint: See the Chapter 19 supplemental diagrams.)

2. A compass needle in London pointed exactly to true north in:
   (a) 1580
   (b) 1660
   (c) 1820
   (d) 1970

(Hint: See the Chapter 19 supplemental diagrams.)

3. Changes in the velocity of seismic waves within the Earth are controlled by:
   (a) Density
   (b) Temperature
   (c) Pressure
   (d) All of the above

(Hint: See the first edition Slide Set.)

4. What two major features are seen most clearly on the geoid of the continental United States?
   (a) San Andreas fault and Appalachians
   (b) Sierra Nevada and Colorado Plateau
   (c) Yellowstone and the Great Lakes
   (d) Everglades and Cascade Range

(Hint: See the second edition Slide Set.)

FINAL REVIEW

After reading this chapter, you should:

- Understand what scientists have learned about Earth's interior from seismic waves
- Understand how they gained that knowledge
- Understand the principle of isostasy
- Know how heat flows from Earth's interior and the effects of the heat flow on geologic processes
- Understand the importance of convection to plate tectonics
- Understand paleomagnetism and its importance

ANSWER KEY

COMPOSITION AND STRUCTURE OF EARTH'S INTERIOR

Crust: 1. Continents; 2. Granitic, basalt, gabbro; 3. Moho, less dense
Mantle: 1. Do pass, are not; 2. Melt (or fluid); 3. Peridotite; 4. Atoms, high pressure
Core: 1. Increase; 2. Iron

PRACTICE MULTIPLE-CHOICE QUESTIONS

1. d, 2. a, 3. b, 4. b, 5. d, 6. d, 7. b, 8. c, 9. a, 10. a, 11. b, 12. c, 13. c

20

Plate Tectonics: The Unifying Theory

The theory of plate tectonics describes the movement of the lithospheric plates and the forces acting between them. It also explains the distribution of many large-scale geologic features—mountain chains, structures on the seafloor, volcanoes, and earthquakes—that result from movements at plate boundaries. Plate tectonics provides the conceptual framework for this book, and indeed for much of geology. This chapter examines the development and implications of plate-tectonic theory.

◆ The initial concept of continental drift was based on the jigsaw-puzzle fit of the coasts on both sides of the Atlantic. At the end of the nineteenth century, geologist Eduard Suess put some of the pieces of the puzzle together and postulated that the combined present-day southern continents had once formed a single giant continent, Gondwanaland. In 1915, Alfred Wegener cited as further evidence the remarkable similarity of rocks, geologic structures, and fossils on opposite sides of the Atlantic. Wegener postulated a supercontinent called Pangaea that began to break up, some 200 million years ago, into the continents as we know them today, with ocean filling the widening gaps between them. In 1928, Arthur Holmes proposed the mechanism of thermal convection in the mantle as the driving force of continental drift.

Convincing evidence began to emerge as a result of extensive exploration of the seafloor after World War II. The discovery of the deep rift running down the center of the Mid-Atlantic Ridge sparked much speculation. In the early 1960s, Harry Hess of Princeton University and Robert Dietz of the University of California suggested that the seafloor separates along the rifts in mid-ocean ridges and that new seafloor forms by the upwelling of hot mantle materials in these cracks, followed by lateral spreading. This is the theory of seafloor spreading.

◆   The present ocean basins are being created by seafloor spreading and consumed by subduction. Geologists have drilled into the floor of the deep ocean in an attempt to find remnants of seafloor older than 200 million years—without success. In 1990, the oldest oceanic rocks were found after a 20-year search in the western Pacific. They were only some 175 million years old. Older oceanic rocks don't exist; they have been subducted.

Continents are more enduring than the seafloor because they are too buoyant to be subducted. They may be fragmented, moved, aggregated, and deformed by the movement of plates, but the pieces survive. The old core of North America, for example, was assembled by plate collisions about 2 billion years ago from pieces of even older continents, some as old as 4 billion years. Continents can be eroded and fragmented, but they can also grow over time by the gradual accumulation of materials along their margins. New continental strips can be added on as plates separate and collide, fragment, move about, and reassemble.

Seafloor spreading is a relatively simple mechanism and is reasonably well understood, but the evolution of continents is enormously complex. Continental rock assemblages, volcanism, metamorphism, and the evolution of mountain chains are being reexamined in the framework of plate tectonics. Most geologists now believe that the geology of continents has been dominated by plate tectonics for at least 3.5 billion years of geologic history.

◆   According to the theory of plate tectonics, the lithosphere is broken into a dozen or so rigid plates that slide over a partially molten, weak asthenosphere. The continents, embedded in some of the moving plates, are carried along passively. Continental drift is a consequence of plate movements.

At divergent boundaries, where plates move apart, partially molten mantle material rises and fills the gap between them. This material becomes new lithosphere added to the trailing edges of the diverging plates. On the seafloor, the boundary between separating plates is marked by a mid-ocean ridge that exhibits active basaltic volcanism, shallow-focus earthquakes, and normal faulting caused by tensional forces created by the pulling apart of two plates. The process by which plates separate and ocean crust is created is called seafloor spreading. Early stages of plate separation can be found on continents. Such sites are characterized by down-faulted rift valleys, basaltic volcanic activity, and shallow-focus earthquakes.

At convergent boundaries, the overridden plate is subducted into the mantle below, where eventually it is recycled. Collision and subduction produce deep-sea trenches, adjacent mountain ranges of folded and faulted rocks, and magmatic belts. The magmatic belt can be a mountain range on land or a chain of volcanic islands, called an island arc, on the seafloor. Once subducted lithosphere is heated, water and other volatiles "boil off." These join hot materials in the wedge of mantle above and induce melting. This is the source of magma that feeds and builds the overlying chain of volcanoes. The collision of two plates generates very large forces in the region, and these forces result in the faulting that triggers the shallow- and deep-focus earthquakes that occur in subduction zones. By the time the lithospheric plate has migrated from a divergent to a convergent boundary, it has cooled to become denser than the mantle below. Some geologists believe that the weight of the sinking part of the plate helps to pull the entire plate down and thus serves as an important part of the driving mechanism of plate tectonics.

At transform boundaries, plates slide past each other, neither creating nor destroying lithosphere. Transform faults occur where the continuity of a divergent boundary is broken and offset. Shallow-focus earthquakes with horizontal slips occur on transform boundaries.

♦ Once it was known that the plates are moving, geologists tried to determine the rate at which they are moving. This was done through two methods: seafloor magnetism and deep-sea drilling. As magma cools at a mid-ocean ridge, the iron in the basalt becomes magnetized by the Earth's magnetic field. The iron points toward magnetic north, and this direction is preserved in the rock as it spreads away from the ridge. Whenever the direction of the Earth's magnetic field reverses polarity (which it does periodically, as we saw in Chapter 19), this change is recorded in the rock being formed at that time. The ages of reversals have been worked out from magnetized lavas on land. Using this known sequence of reversals over time, geologists could assign ages to the bands of magnetized rocks on the seafloor. Because they now knew the age of a band of magnetized rocks on the seafloor and knew the distance from a mid-ocean ridge crest where the magnetized rocks were created, they could calculate the velocity of plate movements. By this method, geologists found that the highest rate of seafloor spreading is 10 to 12 cm per year at the East Pacific Rise, the lowest 2 cm per year at the Mid-Atlantic Ridge.

In 1968, a program began to study seafloor sediments from many places in the world's oceans. Using hollow drills, scientists brought up cores containing sections of seafloor sediments and underlying volcanic crust. Because sedimentation begins as soon as ocean crust forms, the age of the oldest sediments in the core tells us how old the ocean floor is at that spot. It was found that the oldest sediments in the cores become older with increasing distance from mid-ocean ridges and that the age of the seafloor at any one place agrees almost perfectly with the age determined from magnetic reversal data. This agreement validated magnetic dating of the seafloor and clinched the concept of seafloor spreading.

♦ Figure 20.11 shows the ages of the world seafloor as determined from fossil and magnetic reversal data. Each colored band represents a span of time that gives the age of the crust within that band. The boundaries between bands, called isochrons, are contours that connect rocks of equal age. Isochrons show the time that has elapsed, and thus the amount of spreading that has occurred, since the rocks were formed from upwelling magma at mid-ocean ridges.

♦ The fast-moving plates (the Pacific, Nazca, Cocos, and Indian) are being subducted along a large fraction of their boundaries. In contrast, the slow-moving plates (the North and South American, African, Eurasian, and Antarctic) have large continents embedded in them and do not have significant attachments of downgoing slabs. One hypothesis to explain these observations is that rapid plate motions are caused by the pull exerted by large-scale downgoing slabs, and slow plate motions are caused by the drag associated with embedded continents.

♦   The direction of the movement of one plate in relation to another depends on geometric principles that govern the behavior of rigid bodies. Two primary geometric principles govern the movement of plates on our planet: (1) transform boundaries reveal the direction of plate movement; and (2) isochrons reveal the positions of plates in earlier times. By applying these principles, geologists can deduce spreading rates from spreading directions and magnetic anomalies and can work out the history of the motions of all the lithospheric plates. New techniques such as satellite altimetry mapping and Global Positioning System (GPS) measurements have allowed geologists to confirm the results found by magnetic anomaly and isochronic methods.

♦   The only record we have of past geologic events is the incomplete one found in the rocks that have survived erosion or subduction. Because only seafloor younger than 200 million years has survived subduction, we must focus on the continents to find the old rocks that provide the evidence for most of Earth's history. Rock assemblages on the continents that are characteristic of the various kinds of plate boundaries allow geologists to identify ancient episodes of plate separations and collisions.

At ocean-ocean divergent boundaries, ophiolite suites are found. These are rock sequences composed of deep-sea sediments, basaltic pillow lava, gabbro, and peridotites. Ophiolite suites are sometimes found on land, and geologists can now explain these rocks as fragments of oceanic crust that were transported by seafloor spreading and then raised above sea level and thrust onto a continent in an episode of plate collision.

When plate separation begins within a continent, the ancient site of rifting can often be found by another characteristic assemblage of rocks and structures. These include rifts and downdropped blocks of continental crust, volcanic intrusions, and thick sedimentary basins along continental margins. Present-day continental rifting can be seen in the rift valleys of East Africa.

At convergent plate boundaries where two oceanic plates collide, the cooler, denser, heavier plate is subducted below the other one, forming an ocean trench. Water, other volatiles, and melt rise from the heated subducted plate and cause melting of peridotite in the overlying wedge of mantle. Oceanic crust is intruded by magma to form an arc of volcanic islands on the seafloor. A forearc basin fills with arc-derived sediments. An accretionary wedge builds up from ocean sediments and crust scraped off the descending plate. An example of an ocean-ocean plate convergence is the Philippine Islands.

When an oceanic and a continental plate collide, the oceanic plate is subducted below the continental plate, forming a trench. The continental plate buckles, forming mountains; at the same time, the sediments on the oceanic crust are scraped off and added to the continental plate. The Andes Mountains and the Peru-Chile Trench along Chile's coastline are outstanding examples. The jumbled mass of rocks at this type of boundary is called a mélange.

In a collision between continents, neither plate is subducted because continental crust is buoyant. A wide zone of intense deformation develops at the boundary where the continents grind together. This boundary is marked by a mountain range in which sections of deep- and shallow-water sediments are found highly folded and broken by multiple thrust faults. The buildup of thrust sheets results in a much-thickened continental crust in the collision zone. Often, there is a magmatic belt within the mountain range. The remnant of such a zone left behind in the geologic record is called a suture. Ophiolites are often found near the suture; they are relics of an ancient ocean that disappeared in the convergence of two plates. A prime example of a collision of continents is the Himalayas, which began to form some 50 million years ago when a plate carrying India collided with the Eurasian Plate.

At a transform boundary between two oceanic plates, the rocks on either side are of the same type (oceanic crust and sediment) but of different ages and water depths because of differing distances from the mid-ocean ridges where they originated. When a transform fault occurs on land, as in the case of the San Andreas fault, the rocks on either side of the transform boundary are more likely to be of different types and ages. Unlike seafloor rocks, rocks found on land are highly variable from place to place, and a large fault offset is likely to juxtapose rocks that formed under different circumstances of intrusion, metamorphism, deposition, deformation, and erosion.

◆   Microplate terranes are blocks as large as hundreds of kilometers across within continental orogenic belts, with assemblages of rocks that are alien to their surroundings. Fossils indicate that these blocks originated in environments and at times different from those of the surrounding area. Microplate terranes are now believed to be fragments of other continents, seamounts, island volcanic arcs, or slices of ocean crust that were swept up and plastered onto a continent when plates collided or remnants left behind when a continent split apart and separated.

The Appalachian orogenic belt, from Newfoundland to the southeastern United States, contains microplate terranes—slices of ancient Europe, Africa, oceanic islands, and crust welded onto North America during ancient collisions.

◆   Using evidence gathered over the years, scientists have been able to reconstruct the paths of the plates and thus the continents and oceans. *(See the accompanying article on the grand reconstruction.)*

PLATE TECTONICS: THE GRAND RECONSTRUCTION

One of the great triumphs of modern geology is the reconstruction of events that led to the assembly of the supercontinent Pangaea and its later fragmentation into the continents we know today. Pangaea was the only continent existing at the close of the Paleozoic era, some 250 million years ago. It stretched from pole to pole and was made up of smaller continents that collided during the Paleozoic era.

Because the seafloor record of these events has been destroyed by subduction, we must rely on the older evidence preserved on continents to identify and chart the movements of these paleocontinents. Old mountain belts help us locate ancient collisions of the paleocontinents. In many places, alien rock assemblages reveal ancient episodes of rifting and subduction. Rock types and fossils also indicate the distribution of ancient seas, glaciers, lowlands, mountains, and climatic conditions. A knowledge of ancient climates enables geologists to locate the latitudes at which the continental fragments formed, which in turn helps us to assemble the jigsaw fragments of ancient continents. The fossil magnetism of a continental fragment records its ancient orientation and position.

These lines of evidence led scientists to portray the breakup of a supercontinent, Rodinia, 750 million years ago, and to follow its fragments over the next 500 million years as they drifted and reassembled into Pangaea (see Figure 20.23). More is known about the breakup of Pangaea than about the breakup of Rodinia because much of the evidence of Pangaea's breakup is still available on the seafloor.

At the close of the Paleozoic era, Pangaea was an irregularly shaped landmass surrounded by a universal ocean called Panthalassa. This was the ancestral Pacific. The Tethys Sea, between Africa and Eurasia, was the ancestor of part of the Mediterranean. Permian glacial deposits found in widely separated areas are explained by postulating a single continental glacier flowing over the South Polar regions of Gondwanaland in Permian time, before the breakup of the continents.

Figure 20.24 shows the breakup of Pangaea as we now understand it. The breakup of Pangaea began with the opening of rifts from which basalt poured. Rock assemblages that are relics of this great event can be found today in Triassic dikes and sills from Nova Scotia to North Carolina and in the Palisades sill along the Hudson River. Radioactive age dating of these rocks tells us that the breakup and the beginning of drift occurred about 200 million years ago.

After 20 million years of drift, the Atlantic had partially opened, the Tethys had contracted, and the northern continents (Laurasia) had all but split away from the southern continents (Gondwana). New ocean floor also separated Antarctica-Australia from Africa–South America. India was off on a trip to the north.

By the end of the Jurassic period, 140 million years ago, drift had been under way for 60 million years. South America split from Africa, which signaled the birth of the South Atlantic. The North Atlantic and Indian oceans were enlarged, but the Tethys Sea continued to close. India continued its northward journey.

The close of the Cretaceous period, 65 million years ago, saw a widened South Atlantic Ocean, the splitting of Madagascar from Africa, and the close of the Tethys Sea to form the Mediterranean Sea. After 135 million years of drift, the modern configuration of continents becomes discernible.

The modern world was produced over the past 65 million years. India has collided with Asia, ending its trip across the ocean. Australia has separated from Antarctica. Nearly half of the present-day ocean floor was created in this period.

Hardly any branch of geology remains untouched by this grand reconstruction. Economic geologists are using the fit of the continents to find mineral and oil deposits by correlating the formations in which they occur on one continent with their pre-drift continuations on another. Paleontologists are rethinking some aspects of evolution in light of continental drift. Geologists are broadening their focus from regional mapping to the world picture, because the concept of plate tectonics provides a way to interpret such geologic processes as sedimentation and orogeny in global terms. Oceanographers are reconstructing currents as they might have existed in the ancestral oceans to understand the circulation of modern ocean currents better and to account for variations in deep-sea sediments. What better testimony to the triumph of this once outrageous hypothesis than its ability to shed light on so many diverse topics?

♦ We will not fully understand plate tectonics until we can explain why plates move. In Chapter 19, we described the mantle as a hot solid capable of flowing like a liquid at a speed of a few centimeters per year. The plates of the lithosphere somehow move in response to the flow in the underlying mantle. Many hypotheses have been advanced to explain the mechanism of plate motion. Some scientists believe that plates are pushed by the weight of the ridges at the zones of spreading or pulled by the weight of the sinking slab at subduction zones. Others hold that the plates are dragged along by currents in the underlying asthenosphere. Hot spots may spread laterally when they reach the lithosphere, dragging the plates. Downward return flow consists of subducted lithosphere sinking to great depth, possibly as far as the core-mantle boundary.

It is likely that the process is a highly complex convective flow involving rising hot, partially molten materials; a cool surface boundary layer; and sinking cool, solid materials under a variety of conditions ranging from melting to solidification and remelting. A significant part of the mantle must be involved, because seismic tomography shows slabs to penetrate to depths of some 700 km. Among the tasks left to the next generation of Earth scientists is the incorporation of such important details as the shapes of plates, the history of their movements, and the formation and growth of continents into an explanation involving convective currents in the interior.

CHAPTER 20 OUTLINE

TERMS AND CONCEPTS

Match the lettered option listed below the terms to the appropriate term.

1. _____ Continental drift
2. _____ Island arc
3. _____ Isochron
4. _____ Mélange
5. _____ Microplate terranes
6. _____ Pangaea
7. _____ Seafloor spreading
8. _____ Suture

A. A good example of one exists in the Himalayas
B. Associated with convergent ocean-ocean plate boundaries
C. Old concept that underlies plate-tectonics theory
D. Shows the ages of the seafloor
E. Most of Florida could be one
F. Indicates an ancient episode of subduction when paired with magmatism
G. Associated with divergent boundaries
H. "Discovered" by Alfred Wegener in 1915

PLATE MOVEMENTS SINCE TRIASSIC TIME

Using the figures as an aid, complete the statements that follow the figures.

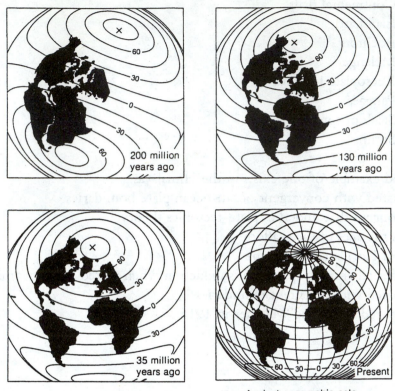

X = Ancient geographic pole

Plate movements have led to the (1) _____ drift of the continents and the opening of the (2) _____ Ocean. The central (3) _____, the (4) _____, and the (5) _____ began to form about (6) _____ million years ago, when (7) _____ began to break up and (8) _____ and (9) _____ drifted away from North America.

     The South Atlantic opened about (10) _____ million years ago with the separation of (11) _____ from (12) _____. As the continents drifted apart, they also migrated in a (13) _____ direction to their present positions.

PRACTICE MULTIPLE-CHOICE QUESTIONS

Circle the option that best answers the question.

1. The idea that the continents move over the globe is called:
   (a) Plate-tectonics theory
   (b) Continental drift
   (c) Breakup of Pangaea
   (d) Jigsaw-puzzle fit of the continents

2. The concept of seafloor spreading was advanced:
   (a) In the early 1960s
   (b) In the mid-1920s
   (c) By Arthur Holmes, who related it to thermal convection in the mantle
   (d) By Alfred Wegener in 1915

3. The oldest rocks on the ocean floor are about:
   (a) 50,000 years old
   (b) 2 million years old
   (c) 200 million years old
   (d) 2 billion years old

4. On the seafloor, the boundary between separating plates is marked by:
   (a) A mid-ocean ridge
   (b) A long downfaulted rift valley
   (c) Deep-focus earthquakes
   (d) Thrust faulting

5. Continental plate separation is marked by:
   (a) Deep-focus earthquakes
   (b) Mountain-building
   (c) Transform faults
   (d) A long downfaulted rift valley

6. Which of the following does not commonly occur along plate convergence boundaries:
   (a) Shallow-focus earthquakes
   (b) Basaltic volcanism
   (c) Andesitic volcanism
   (d) Accumulation of deep-water carbonate deposits

7. Continent-continent plate collisions:
   (a) Are a major producer of folded mountain ranges worldwide
   (b) Are characterized by many deep-focus earthquakes
   (c) Form massive basalt deposits on the surface
   (d) Are best associated with divergent plate boundaries

8. At which of the following boundaries would volcanoes be least likely to form?
   (a) Mid-ocean ridges
   (b) Convergent plate boundaries
   (c) Divergent plate boundaries
   (d) Transform plate boundaries

9. The movement of the Pacific Plate past the North American Plate along the San Andreas fault is an example of what type of plate boundary?
   (a) Continent-continent collision
   (b) Divergent
   (c) Ocean-continent collision
   (d) Transform

10. An area that experiences only shallow-focus earthquakes is:
    (a) The west coast of South America
    (b) Southern Alaska
    (c) The Mid-Atlantic Ridge
    (d) None of these; they all have deep-focus earthquakes

11. All of the following features are associated with plate boundaries except:
    (a) The Peru-Chile Trench along the west coast of South America
    (b) Iceland
    (c) The Hawaiian Islands
    (d) The Himalayas

12. When the magnetic field of the Earth is reversed:
    (a) The Earth flips over in its orbit so that north is always up
    (b) The sense of rotation of the Earth is also reversed
    (c) Magnetization of all existing rocks in the ocean is reversed and match the orientation of the new magnetic field
    (d) Earth's magnetic polarity is such that the north-seeking end of a magnetic compass needle would point toward the south magnetic pole

13. A symmetrical pattern of bands of rocks that are positively and reversely magnetized is called:
    (a) Positive anomaly pattern
    (b) Negative anomaly pattern
    (c) Magnetic anomaly pattern
    (d) Seafloor anomaly pattern

14. The rate of seafloor spreading:
    (a) Is uniform along the entire length of a spreading center
    (b) Cannot be measured
    (c) Is such that the ocean floors contain no rocks older than about 200 million years
    (d) Is greater in deeper water

15. The spreading centers that have the highest rate of movement (the greatest amount of active separation taking place) are located:
    (a) In the North Atlantic
    (b) In the South Atlantic
    (c) Along the East Pacific Rise
    (d) In the southern Indian Ocean

16. Which of the following features are used to reveal the geometry of plate motions?
    (a) Transform boundaries, isochrons
    (b) Divergent boundaries, convergent boundaries
    (c) Divergent boundaries, isochrons
    (d) Convergent boundaries, isochrons

17. Ophiolite suites are created:
    (a) On the ocean floor
    (b) At the edge of a continent
    (c) On land
    (d) Only at convergent plate boundaries

18. Continental shelf deposits are laid down:
    (a) At ocean–ocean divergent boundaries
    (b) At ocean–ocean convergent boundaries
    (c) At receding continental margins
    (d) At mid–ocean ridges

19. A jumbled mixture of high-pressure, low-temperature metamorphic rocks found at ocean-continent convergences is called:
    (a) An ophiolite suite
    (b) A mélange
    (c) A microplate terrane
    (d) Metamorphosed sediments

20. Assemblages of smaller, "foreign" pieces of continental material that have been tacked onto a continent's core are called:
    (a) Microplate terranes
    (b) Mélanges
    (c) Ophiolite suites
    (d) Island arcs

21. The oceans that existed at the end of the Paleozoic era were:
    (a) Atlantic and Pacific
    (b) Atlantic and Panthalassa
    (c) Atlantic and Tethys Sea
    (d) Panthalassa and Tethys Sea

CD RESOURCES

SUPPLEMENTAL DIAGRAMS

20.26 Principal features of mid-ocean divergent plate margin
20.27 Comparison of igneous rock ages at different distances from mid-ocean ridges
20.28 Map of the world's ocean ridges
20.29 Worldwide distribution of ophiolites
20.30 Sediment accumulation along passive margin of the eastern United States since the Jurassic
20.31 Geology of the western United States at the beginning of Tertiary time
20.32 Terranes of the North Pacific
20.33 Terranes of the South Pacific
20.34 Wrangellia terrane (diagram to accompany supplemental photo 20.41)
20.35 Chulitna terrane (diagram to accompany supplemental photo 20.42)
20.36 Panthalassa
20.37 Reconstruction of the ancient continent of Pangaea
20.38 Whole Earth maps illustrating breakup of Pangaea
20.39 All of the world's identified hot spots

SUPPLEMENTAL PHOTOS

20.40 Landsat image of six terranes, coast of southeastern Alaska
20.41 Wrangellia terrane (see supplemental diagram 20.34)
20.42 Chulitna terrane (see supplemental diagram 20.35)

FIRST EDITION SLIDE SET

Slide 20.1 Age of the ocean basins
Slide 20.2 Gulf of California, southwestern North America
Slide 20.3 The Nile delta, Sinai Peninsula, and Red Sea
Slide 20.4 Farallon de Pajaros (Uracus), Mariana Islands, Pacific
Slide 20.5 Spreading center in a lava pool, Kilauea's east rift, Hawaii
Slide 20.6 Zone of crustal convergence in a lava pool, Kilauea's east rift, Hawaii

SECOND EDITION SLIDE SET

See the slides for Chapter 21.

EXPANSION MODULES

Chapter 31 The Himalayas and Tibetan Plateau

ANIMATIONS

20.1   Plate Tectonics: Plate Divergence (Red Sea)
20.2   Plate Tectonics: Continent–Continent Convergence (Himalayas)
20.3   Plate Tectonics: Ocean–Continent Convergence (Andes)
20.4   Plate Tectonics: Transform Boundary (San Andreas)
20.5   Plate Tectonics: Passive Continental Margin

ILLUSTRATED GLOSSARY

EXERCISES

World Plate Boundaries
Geology of Plate Boundaries

CD LAB

1. Which of the world's ocean ridges is currently undergoing a fast rate of separation?
    (a) Mid-Atlantic Ridge
    (b) East Pacific Rise
    (c) Southeast Indian Ridge
    (d) Atlantic-Indian Ridge

    (Hint: See the Chapter 20 supplemental diagrams.)

2. The ophiolites running along the Appalachians mark the closing of:
    (a) Panthalassa
    (b) The Tethys Ocean
    (c) The Iapetus Ocean
    (d) The Atlantic Ocean

    (Hint: See the Chapter 20 supplemental diagrams.)

3. The Gulf of California began to form _____ years ago, as Baja separated from mainland Mexico:
    (a) 4 to 6 million
    (b) 10 to 20 million
    (c) 100 to 200 million
    (d) About 1 billion

(Hint: See the first edition Slide Set.)

4. The average life of a volcano in an island arc system is about:
    (a) 1 million years
    (b) 2 million years
    (c) 3 million years
    (d) 4 million years

(Hint: See the first edition Slide Set.)

5. The Nazca Plate is moving toward the South American Plate at a rate of about:
    (a) 2 cm/year
    (b) 4 cm/year
    (c) 6 cm/year
    (d) 8 cm/year

(Hint: See the second edition Slide Set for Chapter 21.)

6. What type of faults are found near the continental edge where the African Plate and the Arabian Plate are diverging to form the Red Sea?
    (a) Normal
    (b) Reverse
    (c) Strike-slip
    (d) Thrust

(Hint: See Animation 20.1.)

7. At the convergent plate boundary where the Himalayas are being uplifted by the collision of India with Asia, what type of faults are found where material is scraped off the lower plate?
    (a) Normal
    (b) Reverse
    (c) Strike-slip
    (d) Thrust

(Hint: See Animation 20.2.)

Web Project: Among the many links on the Chapter 20 page, a standout is *This Dynamic Earth: The Story of Plate Tectonics*, USGS, an on-line edition of a book written by W. Jacquelyne Kious and Robert I. Tilling. If you're interested in learning more about plate tectonics, this is a very good place to start. Illustrated with many easy-to-read maps and diagrams, as well as photos, the online book has chapters on the history of plate tectonics; the theory of plate tectonics; plate motions; unanswered questions about plate tectonics (such as the driving mechanism and how plate tectonics may operate on other planets); and people and plate tectonics, including sections on earthquakes, volcanoes, and geothermal energy. This book is an excellent overview of plate tectonics and is well worth checking out.

FINAL REVIEW

After reading this chapter, you should:

- Know the text's description of the gradual acceptance of the plate-tectonics theory and understand what the story suggests about scientific progress
- Have a clear understanding of the theory of plate tectonics
- Understand the importance of the seafloor in the theory of plate tectonics
- Know the geologic characteristics of plate boundaries and the rock assemblages found there
- Understand the basics of how the present-day positions of the continents evolved over the past 200 million years
- Understand plate motion and the theory that is offered to explain the driving mechanism behind it

ANSWER KEY

TERMS AND CONCEPTS

1. C, 2. B, 3. D, 4. F, 5. E, 6. H, 7. G, 8. A

PLATE MOVEMENTS SINCE TRIASSIC TIME

1. Northward, 2. Atlantic, 3. Atlantic, 4. Caribbean, 5. Gulf of Mexico, 6. 200, 7. Pangaea, 8. Africa, 9. South America, 10. 150, 11. South America, 12. Africa, 13. Northward

PRACTICE MULTIPLE-CHOICE QUESTIONS

1. b, 2. a, 3. c, 4. a, 5. d, 6. d, 7. a, 8. d, 9. d, 10. c, 11. c, 12. d, 13. c, 14. c, 15. c, 16. a, 17. a, 18. c, 19. b, 20. a, 21. d

21

Deformation of the Continental Crust

As we saw in Chapter 20, most of Earth's surface consists of oceans whose underlying crust is younger than 200 million years. Only the continents, which cover less than a third of Earth's surface, have rocks as old as 4 billion years, almost as old as Earth itself, and it is there that we must turn to look back over most of geologic time.

This chapter examines some vertical movements of the continental crust over the past 4 billion years. The rocks that make up the continental crust can be grouped into two categories: undeformed sedimentary rocks, and rocks of all three types that have been deformed and altered by crustal forces. Most of the continental crust consists of rocks in the second category. Orogeny—the mountain-building processes of folding, faulting, magmatism, and metamorphism—is intimately related to the evolution of continents.

♦ The geology of continents exhibits a pattern of eroded remnants of very old deformed rocks in the interior and more recent deformation in the mountain systems closer to the margins. Mountain building occurs where plates collide. A subducted plate melts, and magma rises into the deformed belt. Plate movements bring in foreign fragments and weld them to the deformed belt. Upward and downward movements within a continent create interior basins and domes and lift old mountains high again. Offshore, downward movements create basins on the continental shelves. Figure 21.2 is a map showing the world pattern of deformed continental rocks, colored according to the geologic period in which the deformation occurred.

♦ Cratons are the extensive, flat, tectonically stable interiors of the continents. They are composed of ancient rocks that underwent intense deformation in Precambrian time but have been relatively quiescent since then. Cratons usually include large shields that consist of very old, exposed crystalline basement rocks. From the shields come the oldest rocks ever found on Earth. Shields are found in Canada, Scandinavia and Finland, Siberia, central Africa, Brazil, and Australia.

On the North American continent, the Canadian Shield forms the core of the continent. Surrounding it is a sediment-covered, almost level region called a platform, which is a subsurface continuation of the Canadian Shield. The platform is covered with a thin veneer of Paleozoic sedimentary rocks. Sedimentary basins, broad depressions where sediments are thicker than the surrounding rock, are often found within platforms.

◆   Around the stable interior are orogenic belts, areas of younger deformation and current tectonic activity. During an orogeny, continental rock is folded and faulted, broken up, and deformed to create mountains. (Figure 21.12 depicts some of the major types of mountains found in orogenic belts.) When two continents collide, the crust can break into thrust sheets where large portions of the crust are thrust over other portions. These thrust sheets are often deformed and metamorphosed. A significant fraction of an orogenic belt can be made up of a succession of displaced or microplate terranes, fragments of crust that may have traveled thousands of kilometers from distant parts of the world before they were added to the orogenic belt. Continental crust can also be thickened when plutonic and volcanic rocks derived from melting of the mantle are added.

◆   The old, eroded Appalachian Mountains are a classical fold-and-thrust belt consisting of several distinct provinces. In the Valley and Ridge province, thick Paleozoic sedimentary rocks laid down on an ancient continental shelf were folded and thrust to the northwest by compressional forces from the southeast. The eroded mountains of the Blue Ridge province are composed largely of Precambrian and Cambrian crystalline rock, showing much metamorphism. The Blue Ridge rocks were not intruded and metamorphosed in place but were thrust as sheets over the sedimentary rocks of the Valley and Ridge province. The Piedmont province contains Precambrian and Paleozoic metamorphosed sedimentary and volcanic rocks intruded by granite, all now eroded to low relief. The Piedmont was thrust over Blue Ridge rocks along a major thrust fault, overriding them to the northwest. In the Coastal Plain province, relatively undisturbed sediments of Jurassic age and younger are underlain by rocks similar to those of the Piedmont. The continental shelf is the offshore extension of the Coastal Plain. Figure 21.9 shows a possible plate-tectonic reconstruction of southern Appalachian history consistent with present-day geologic features.

◆   The stable interior platform of North America is bounded on the west by the North American Cordillera, a mountain belt that contains some of the highest peaks on the continent.. Across its middle section, the Cordilleran system is about 1600 km wide and includes several contrasting physiographic provinces: the Coast

Ranges along the Pacific Ocean; the Sierra Nevada and Cascades; the Basin and Range province (a region of faulted and tilted blocks that form many narrow mountain ranges and valleys); the high tableland of the Colorado Plateau; and the Rocky Mountains, which end abruptly at the edge of the Great Plains.

The Cordilleran system is topographically higher and more extensive than the Appalachians because its main orogeny was more recent (the last half of the Mesozoic era and early Tertiary period). The form and height of the Cordillera today are also manifestations of even more recent events over the last 15 or 20 million years, which resulted in rejuvenation of the mountains. Rejuvenation means that the mountains were raised again and brought back to a more youthful stage. At that time, the central and southern Rockies attained much of their present height as a result of a broad regional upwarp. Other examples of rejuvenation by upwarping are the Adirondacks of New York, the Labrador Highlands, and the mountains of Scandinavia and Finland. Rejuvenated deformed belts are also seen in the modern topography of the Alps, the Urals, and the Appalachians. The cause of rejuvenation is a matter of t debate among geologists.

The late structural imprinting responsible for the present-day features of the Basin and Range province and the tilted uplift of the Sierra Nevada of California occurred during a spectacular episode of faulting over a large region. Thousands of nearly vertical faults sliced the crust into innumerable upheaved and downdropped blocks, forming hundreds of nearly parallel fault-block mountain ranges separated by rift valleys bounded by normal faults.

The Colorado Plateau seems to be an island of the central stable region, cut off from the interior by the Rocky Mountains. Since the late Precambrian, it has been a stable area. Its rock formations, exposed in the Grand Canyon, reflect mainly up-and-down movements.

♦   The Atlantic coastal plain and the continental shelf began to develop in the Triassic period with the rifting that preceded the opening of the modern Atlantic Ocean. The rift valleys formed basins that trapped a thick series of nonmarine sediments. As these deposits were accumulating, they were intruded by basaltic sills and dikes. The coastal plain and shelf of the Gulf of Mexico are continuous extensions of the Atlantic coastal plain and shelf, interrupted only briefly by the Florida peninsula.

♦   Epeirogeny is another kind of crustal movement, the gradual downward and upward movements of the crust without significant deformation. Although many of the vertical movements are connected with orogeny, slow, intermittent epeirogenic movements commonly affect large regions without producing extensive folding or faulting. A typical product of epeirogenic downward movement is a basin in the stable interior, such as the Michigan Basin. Upward movement with little or only moderate faulting or folding can produce broad uplands and plateaus. The Colorado Plateau and the Black Hills of South Dakota were both produced by such upward movements. Geologists have no definite explanation for most of these slow and broad movements, but they have hypotheses that may account for some of them. The uplift of Scandinavia and Finland and the raised beaches of northern Canada represent the slow upward recovery of the crust after the removal of the glacial load that had depressed it. The formation of deep basins on both sides of mid-ocean ridges is believed to be caused by the cooling and contraction of the newly formed ocean plate. Heating of the lithosphere from below can result in thinning and upwarping. Movements in the mantle can stretch and thin the lithosphere above, without breaking the plate. This stretching and thinning may explain the subsidence of basins within continents. Intrusion of magma can thicken the continental crust and result in uplift. In general, however, warping within continents or anywhere far from plate margins is not completely understood.

CHAPTER 21 OUTLINE

MOUNTAIN STRUCTURES

Match the examples to the correct type of mountain structure.

A. Volcanic _____
B. Upwarped _____
C. Fault block _____
D. Folded _____
E. Stacked _____

1. Adirondacks
2. Alps
3. Andes
4. Appalachians – Southern
5. Appalachians – Valley and Ridge
6. Cascades
7. Front Range of Rocky Mountains
8. Himalayas
9. Sierra Nevada
10. Teton Range
11. Urals
12. Wasatch Range

PRACTICE MULTIPLE-CHOICE QUESTIONS

Circle the option that best answers the question.

1. The geology of continents follows the general pattern:
   - (a) Old, eroded mountain ranges in the interior, surrounded by a stable craton
   - (b) Old deformed and eroded rocks in the stable interior, with young mountain ranges near the continental margins
   - (c) Young mountain ranges in the interior, with basins at the continental margins
   - (d) Old deformed and eroded rocks in the stable interior, with older eroded mountain ranges near the continental margins

2. Which of the following statements is true?
   - (a) A continental platform generally includes a craton
   - (b) A shield generally includes a craton
   - (c) A craton generally includes a shield, surrounded by a platform
   - (d) The platform includes the continental margins

3. A shield typically contains:
   - (a) Large deposits of undeformed sedimentary rock
   - (b) Granitic and metamorphic rocks along with deformed sediments
   - (c) Freshly formed volcanic deposits
   - (d) Thick sequences of marine sediments

4. The rocks on the North American platform indicate sedimentation related to:
   - (a) Unsorted deposits associated with the uplift of mountain ranges
   - (b) Abyssal plain deposits associated with deep oceans
   - (c) Extensive shallow inland seas
   - (d) Restricted basins similar to the Mediterranean Sea of today

5. Orogenic activity includes:
   - (a) Faulting, folding, metamorphism, and magmatism
   - (b) A worldwide rise in sea level
   - (c) Reversals of the magnetic field
   - (d) The shrinking of shields and cratons

6. Orogenic activity is most likely associated with:
   - (a) Divergent plate boundaries
   - (b) Activity occurring near a mid-ocean ridge or spreading center
   - (c) The relatively quiet processes found within a shield or craton
   - (d) Plate collisions that occur along convergent boundaries

7. Which of the following provinces is *not* found in the Appalachian Mountains?
   (a) Basin and Range
   (b) Blue Ridge
   (c) Valley and Ridge
   (d) Piedmont

8. The melting of glacial ice has a direct effect on the:
   (a) Loss of seawater
   (b) Upwarping and uplift of large expanses of continental regions
   (c) Rate of seafloor spreading at a mid-ocean ridge
   (d) Formation of metamorphic deposits at a plate boundary

9. The formation of the Himalayas, the world's highest mountains, is explained by:
   (a) A hot spot located near the northern border of India with Tibet
   (b) Epeirogenic activity occurring in the region
   (c) Subduction of an oceanic plate beneath Asia
   (d) The collision of two continental masses

10. Epeirogenic movements can result from:
    (a) Worldwide changes in sea level
    (b) Isostatic readjustments to maintain gravitational equilibrium
    (c) Reversals of the Earth's magnetic field
    (d) Relative movements of the asthenosphere

11. Evidence for epeirogenic movements is found in:
    (a) Metamorphosed rocks at the edges of plates
    (b) Thick sequences of volcanic deposits
    (c) The exposure of large plutonic bodies
    (d) Sequences of relatively undisturbed sedimentary layers

12. Extension of the lithosphere is a reasonable hypothesis to explain:
    (a) A worldwide change in sea level
    (b) Orogenic activity occurring along plate margins
    (c) Subsidence of intracratonic basins, such as the Michigan Basin
    (d) Epeirogenic activity

13. An area where orogenic activity is occurring today is:
    (a) Along the west coast of South America
    (b) In the Midwest of the United States
    (c) In the middle of the Canadian Shield
    (d) Along the east coast of the United States

CD RESOURCES

SUPPLEMENTAL DIAGRAMS

21.22 Map of Appalachians from Alabama to Newfoundland
21.23 Digital shaded-relief map of Appalachian provinces
21.24 Southern Appalachian provinces
21.25 Idealized mountain chain at ocean-continent convergent plate boundary
21.26 Rates of uplift and subsidence in the United States today

SUPPLEMENTAL PHOTOS

21.27 *Challenger* view of the Himalayas and Tibet
21.28 Satellite image of the southern Appalachians
21.29 The Blue Ridge Mountains of Virginia, Blue Ridge province
21.30 Landsat image of Basin and Range region of Nevada
21.31 Tilted and folded sedimentary rocks in the Andes

FIRST EDITION SLIDE SET

Slide 21.1 Mount McKinley, Alaska Range
Slide 21.2 The Valley and Ridge province of the Appalachian Mountains
Slide 21.3 Coast and Andes Mountains, Chile

SECOND EDITION SLIDE SET

Slide 21.1 Housing tracts constructed in recent years along the San Andreas fault zone, San Francisco Peninsula, California
Slide 21.2 Landsat Satellite Thematic Mapper Image of the Los Angeles Area, California
Slide 21.3 Aerial view of offset stream gullies due to right-lateral strike slip along the San Andreas fault, Carrizo Plain, California
Slide 21.4 Aerial view along the San Andreas fault, crossing the Carrizo Plain, California
Slide 21.5 Orange grove offset along a trace of the San Andreas fault system during the 1940 Imperial Valley earthquake, southern California
Slide 21.6 Alaska Range and Mount McKinley
Slide 21.7 Aerial view of the Muldrow Glacier, Alaska Range
Slide 21.8 Indian-Eurasian collision system from space
Slide 21.9 The Himalayan front and Tibetan Plateau from the Space Shuttle
Slide 21.10 Mount Everest (8850 m; 29,028 ft) viewed from near its base in Tibet
Slide 21.11 Large-scale syncline in the Indus Suture Zone, southern Tibet
Slide 21.12 Saipal Himalayas and Lesser Himalayas in western Nepal
Slide 21.13 Middle Siwaliks at Surai Khola, Nepal
Slide 21.14 Bouldery stream channel draining the High Himalayas in Nepal

Slide 21.15 South America from space
Slide 21.16 Global perspective of the Nazca and South American plates
Slide 21.17 Subducting Nazca slab and Andean topography
Slide 21.18 Atacoma Desert in northern Chile
Slide 21.19 View from a seismic station in the Western Cordillera, westernmost Bolivia
Slide 21.20 Dry Salar de Uyuni in the Altiplano of western Bolivia
Slide 21.21 Wet Salar de Uyuni, Bolivia, after a storm
Slide 21.22 A seismic station in the Eastern Cordillera, Bolivia
Slide 21.23 Instrumentation vault of a BANJO seismic station, Bolivia
Slide 21.24 Maragua Syncline, a nappe in the Eastern Cordillera of Bolivia
Slide 21.25 Seismic station and view of the Subandean Zone, Bolivia
Slide 21.26 The Subandean Zone fold and thrust belt, Bolivia
Slide 21.27 On the western edge of the Amazon Jungle, Bolivia
Slide 21.28 Cotopoxi, a composite volcano, Ecuador
Slide 21.29 Fresh lava, Fernandina Island, Galapagos, Ecuador
Slide 21.30 Red Sea divides to form the Gulf of Suez (left) and the Gulf of Aqaba (right)
Slide 21.31 The Red Sea Rift between the African and Arabian plates
Slide 21.32 Cinder Cone and recent lava flow, Lake Turkana, East Africa Rift, Kenya
Slide 21.33 Faulted volcanic ridge, East Africa Rift, Kenya
Slide 21.34 Geysers, Lake Bogoria, East Africa Rift, Kenya
Slide 21.35 Coring sediments in Lake Tanganyika, East Africa Rift, between Tanzania and Zaire
Slide 21.36 Thingvellir graben (aerial view northeast) on the Mid-Atlantic Ridge, southwestern Iceland
Slide 21.37 Thingvellir graben (aerial view southwest) on the Mid-Atlantic Ridge, southwestern Iceland
Slide 21.38 En echelon fractures (fissures) along the northwest shoreline of Lake Thingvellir, southwest Iceland
Slide 21.39 Pillow lavas on the Mid-Atlantic Ridge, Reykjanes Peninsula, Iceland
Slide 21.40 Krafla volcanic complex in northeastern Iceland
Slide 21.41 Laki fissure, Grimsvotn Rift Zone in south central Iceland
Slide 21.42 Skjaldbreidur basaltic shield volcano, Thingvellir graben, southwestern Iceland
Slide 21.43 The Domadalshraun mixed-magma lava flow, Torfajökull volcanic complex, south central Iceland
Slide 21.44 Tephra chronology exposed at the Torfajökull volcanic complex, south central Iceland
Slide 21.45 Archeological excavation at the 12th century Stong Farm Viking settlement, Iceland
Slide 21.46 The harbor at Heimaey, Vestmannaeyjar Islands, Iceland
Slide 21.47 Composite satellite image of western United States
Slide 21.48 Dike swarm in the Canadian Shield

EXPANSION MODULES

Chapter 31 The Himalayas and Tibetan Plateau

ANIMATIONS

20.2   Plate Tectonics: Continent–Continent Convergence (Himalayas)

ILLUSTRATED GLOSSARY

EXERCISES

Structural Provinces of North American Cordillera

CD LAB

1. The Carolina Slate Belt is part of which Appalachian province?
   (a) Blue Ridge
   (b) Valley and Ridge
   (c) Coastal Plain
   (d) Piedmont

   (Hint: See the Chapter 21 supplemental diagrams.)

2. The Great Smoky Mountains are part of which Appalachian province?
   (a) Blue Ridge
   (b) Valley and Ridge
   (c) Coastal Plain
   (d) Piedmont

   (Hint: See the Chapter 21 supplemental photos.)

3. Mount McKinley is part of which fault–thrust system?
   (a) San Andreas
   (b) San Gabriel
   (c) Benioff Zone
   (d) Denali

   (Hint: See the first edition Slide Set.)

4. The Eastern Cordillera is found mostly in:
   (a) Brazil
   (b) Bolivia
   (c) Argentina
   (d) Peru

   (Hint: See the second edition Slide Set.)

5. Four of the thirteen highest peaks in the world are in the:
   (a) Himalayas
   (b) Andes
   (c) Rocky Mountains
   (d) Karakorum Range

(Hint: See Expansion Module Chapter 31.)

FINAL REVIEW

After reading this chapter, you should:

* Know what orogeny is and what orogenic belts are
* Know something about the major belts of deformation in North America
* Understand the relationship between orogeny and plate convergence
* Know what epeirogeny is and how it differs from orogeny

ANSWER KEY

MOUNTAIN STRUCTURES

A. 3, 6   B. 1, 2, 7, 11   C. 9, 10, 12   D. 5 E. 4, 8

PRACTICE MULTIPLE-CHOICE QUESTIONS

1. b, 2. c, 3. b, 4. c, 5. a, 6. d, 7. a, 8. b, 9. d, 10. b, 11. d, 12. c, 13. a

22

Energy Resources from the Earth

This chapter looks at the energy resources we take from the Earth to power our civilization. It covers how these fuels form, where they are found, who controls them, how long supplies will last, and our alternatives when they are exhausted. Resources include all the materials potentially available now and in the future, assuming that changing technology and economics give us new ways to gain access to these materials. Reserves are deposits that have already been discovered and that can be mined economically and legally at the present time.

◆   Figure 22.3 summarizes the history of energy use in the United States. Today, oil, coal, and natural gas supply about 90 percent of the energy used in the United States. All these fuels are derived from organic materials. If we burn wood that was buried and transformed into coal 300 million years ago, we are using energy stored by photosynthesis from late Paleozoic sunlight. We are thus recovering "fossilized" energy. Crude oil and natural gas were also created by a process of burial and chemical transformation of dead organic matter. All such resources derived from natural organic materials are called fossil fuels.

◆   Crude oil (petroleum) and natural gas begin to form when more organic matter is produced than is destroyed by scavengers and decay. This condition exists in environments where the production of organic matter is high—as in the coastal waters of the sea, where large numbers of organisms thrive—and the supply of oxygen in bottom sediments is inadequate to decompose all organic matter. Many offshore sedimentary basins on continental shelves satisfy both of these conditions.

During millions of years of burial, chemical reactions slowly transform some of the organic material into liquid and gaseous compounds of hydrogen and carbon (hydrocarbons). The low density of oil and gas causes them to rise to the highest place they can reach, where they float atop the water that almost always occupies the pores of permeable formations. Geological conditions that favor the large-scale accumulation of oil and natural gas are combinations of structure and rock types that create an oil trap, an impermeable barrier to upward migration. Figure 22.5 shows several different types of oil traps. Not every oil trap contains oil or gas. A trap will contain oil only if source beds were present, if the necessary chemical reactions took place, and if the oil could migrate into the trap and stay there without being disturbed by subsequent heating or deformation. Although oil and gas are not rare, most of the easy-to-find deposits have already been located, and new fields are becoming more difficult to find.

Continental shelves are good places to explore for oil, but offshore drilling poses the threat of environmental pollution. The environment around Santa Barbara, California, for example, suffered great damage in 1969 when oil was accidentally released from an offshore drilling platform. The grounding of the tanker *Exxon Valdez* off the coast of Alaska in 1989 released 240,000 barrels of crude oil in pristine coastal waters, causing severe ecological damage.

We must recognize that oil is indeed exhaustible. For several decades, everyone thought that worldwide reserves would last forever. Analysts now realize that this is not the case and that we must begin to conserve and plan for alternative energy sources for the future. That twice as much oil was removed from the ground in the past 20 years as in the previous 100 years should make us think about the next 10 to 20 years. In addition to the fact that world oil supplies may last for only another 85 years, the price of oil will surely escalate until oil use slows considerably. At the current rate of use, remaining oil reserves in the United States constitute roughly a 20-year supply. Twenty years is not much time before the United States would become dependent on imports for most of its oil, at a cost of perhaps several hundred billion dollars annually.

Possible solutions include the development of new technology to find oil deposits that have been missed by current methods of exploration and to enhance the recovery of oil from existing domestic oil fields; the development of engines with much greater fuel efficiency; the use of alternative fuel sources such as alcohol and natural gas; and conservation.

The resources of natural gas are comparable to those of crude oil and may exceed them in the decades ahead. The world's resources of natural gas are less depleted than oil resources because natural gas is a relative newcomer on the energy scene. The burning of natural gas releases less carbon dioxide per unit of energy than the combustion of coal or oil and is therefore less polluting.

♦   Coal is produced by the burial and chemical transformation of large accumulations of plant material. The first step is the transformation of organic matter to peat, such as in swamps and peat bogs. Over time, with continued burial, the peat is compressed and heated. Chemical transformations increase the peat's already high carbon content, and it becomes lignite, a very soft brownish-black coal-like material containing about 70 percent carbon. The higher temperatures and structural deformation that accompany greater depths of burial may metamorphose the lignite into subbituminous and bituminous coal, or soft coal, and ultimately to anthracite, or hard coal. The greater the metamorphism, the harder and brighter the coal and the higher its carbon content, which increases its heat value. Anthracite is over 90 percent carbon.

Domestic coal resources in the United States would last for 300 to 400 years at current rates of use. Coal has supplied an increasing proportion of the energy needs of the United States since 1975, when the price of oil began to rise, and currently accounts for about 24 percent of the energy consumed. There are serious environmental problems with the recovery and use of coal, however, such as acid rain, toxic coal ash, and strip mining.

♦   Nuclear energy depends on reserves of uranium. Although uranium is present in very small amounts in Earth's crust, it is potentially our largest minable energy resource in terms of energy content (see Figure 22.9). At one time, nuclear energy was promoted as the cheap, clean alternative to fossil fuels. The costs of building and maintaining reactors have since proved prohibitive. There are also serious environmental problems with nuclear energy use, as shown by the accidents at Chernobyl in 1986 and Three Mile Island in 1979. In addition, the uranium used up in nuclear reactors leaves behind dangerous radioactive wastes that must be disposed of. A system of safe long-term waste disposal is not yet available, and reactor wastes are being held in temporary storage at reactor sites.

♦   In principle, the Sun can provide us with all the energy we need, in all the forms we use. Light from the Sun can be converted to heat and electricity. It can even be used to obtain hydrogen, which can be used as a gaseous fuel, from water. Solar energy is risk-free and nondepletable—the Sun will continue to shine for at least the next several billion years. Unfortunately, the technology currently available for large-scale conversion of solar energy to useful forms is inefficient and expensive, although the situation is improving.

In addition to the direct conversion of solar energy into electricity, there are three other energy sources related to the Sun: biomass, hydroelectric energy, and wind power. Solar energy is stored in biomass—material of biological origin, such as wood, grain, sugar, and municipal wastes. Energy can be extracted by direct combustion or by processes that convert biomass to gaseous or liquid fuels, such as methane and alcohol. Hydroelectric energy is derived from water that falls by the force of gravity and is made to drive electric turbines. Hydroelectric energy is clean, relatively riskless, and cheap. Significant expansion of the present capacity would be resisted in the United States, however, because it would involve the drowning of farmlands and wilderness areas under reservoirs behind dams. The use of windmills to drive an electric generator is slowly growing in some places as designs improve and costs are brought down.

◆   Geothermal energy is produced when underground heat is transferred by water that is heated as it passes through a subsurface region of hot rocks that may be hundreds or thousands of feet deep. The water is brought to the surface as hot water or steam through boreholes drilled for the purpose. Eighteen countries now generate electricity using geothermal heat. Reykjavík, the capital of Iceland, is entirely heated by geothermal energy derived from volcanic heat. Like most of the other energy sources we have looked at, geothermal energy presents some environmental problems. Regional subsidence can occur if hot groundwater is withdrawn without being replaced. In addition, geothermally heated waters can contain salts and toxic materials dissolved from the hot rock. These waters present a disposal problem if they are not reinjected.

◆   Using energy more efficiently is like discovering a new source of fuel. It has been calculated that since the rise in oil prices in 1973 turned us all into conservationists, the world has saved more energy than it has gained from all new sources discovered in the same time. Some experts believe that conservation alone could cut the energy used by the industrialized nations in half. The kinds of practices that could lead to these savings require the application of mostly familiar technologies: changes from incandescent to fluorescent lighting; better home insulation; more efficient refrigerators, air conditioners, furnaces, and other appliances; more efficient motors, pumps, and other industrial devices; and better-performing automobile engines.

CHAPTER 22 OUTLINE

ENERGY TERMS AND CONCEPTS

Fill in the blanks with the appropriate word or term that completes the sentence.

1. The energy source that underwent the greatest expansion during the last quarter of the twentieth century was _____ energy.

2. Sedimentary rock that contains a large amount of organic matter from which oil can be extracted is called _____ _____.

3. In the mid-1800s, the leading source of energy used in the United States was _____.

4. Since 1925, the energy source that has shown the largest decrease in usage in the United States is _____.

5. Since 1925, the energy source that has shown the largest increase in usage in the United States is _____.

6. The arched feature consisting of a permeable sandstone layer overlain by an impermeable layer that collects hydrocarbons is a(n) _____ _____.

7. In North America, the largest concentration of tar sands are found in _____, _____.

8. The ultimate energy source is _____.

9. _____ _____ uses the natural heating of water by subsurface heat sources, such as magma chambers.

PRACTICE MULTIPLE-CHOICE QUESTIONS

Circle the option that best answers the question.

1. If an impervious shale overlies a porous oil-bearing sandstone, the structure that could not form an oil trap is:
   (a) An anticline
   (b) A syncline
   (c) A normal fault in dipping strata
   (d) A reverse fault in dipping strata

2. The number of U.S. gallons contained in a barrel of oil is:
   (a) 16
   (b) 25
   (c) 42
   (d) 55

3. The richest oil fields on Earth are located:
   (a) In the Middle East
   (b) On the north slope of Alaska
   (c) On the continental shelf off the coast of West Africa
   (d) In the Caribbean and Gulf of Mexico

4. The United States ranks _____ in oil reserves:
   (a) First
   (b) Second
   (c) Fifth
   (d) Eighth

5. The energy source that has the potential to produce the most energy is:
   (a) Uranium oxide
   (b) Crude oil
   (c) Natural gas
   (d) Coal

6. Which country has the greatest coal reserves?
   (a) China
   (b) Great Britain
   (c) The former Soviet Union
   (d) The United States

7. The major user of natural gas in the United States is:
   (a) The generation of electric power
   (b) Industrial and commercial enterprises
   (c) Residential users
   (d) All three are roughly equal in their use of natural gas

8. The country that has the most syncrude is:
    (a) China
    (b) Great Britain
    (c) The former Soviet Union
    (d) The United States

9. Acid rain results from the combustion of:
    (a) Coal
    (b) Natural gas
    (c) Oil
    (d) Hydrocarbons in general

10. What percentage of electric energy does the United States produce from nuclear power?
    (a) 10
    (b) 20
    (c) 30
    (d) 50

11. The fossil fuel that constitutes the largest potential source of energy for the future is:
    (a) Coal
    (b) Oil and gas
    (c) Oil shale
    (d) Tar sands

12. Which of the following natural processes is the largest potential source of energy for the future?
    (a) Heat flow from the interior of the Earth
    (b) Downhill movement of water
    (c) Tidal movement in the oceans
    (d) Solar radiation arriving at Earth's surface

13. The source that produces the least amount of energy is:
    (a) Hydroelectric
    (b) Wind
    (c) Geothermal
    (d) Nuclear

14. Perhaps the best solution for potential energy shortages in the future is:
    (a) Increasing our energy conservation
    (b) Using more efficient gasoline engines
    (c) Developing more hydroelectric plants
    (d) Discovering new oil fields

CD RESOURCES

SUPPLEMENTAL DIAGRAMS

22.18 Structure of the hydrocarbon butane
22.19 Age distribution of world's past oil production and reserves
22.20 How a nuclear power plant works

SUPPLEMENTAL PHOTOS

22.21 Peat bed in Pleistocene clay bank.

SECOND EDITION SLIDE SET

See Slide 10.7.

ILLUSTRATED GLOSSARY

EXERCISES

Fossil Fuel Cycle

CD LAB

1. More than 80 percent of the world's oil was formed in the last _____ percent
    of geological time:
    (a) 50
    (b) 25
    (c) 12
    (d) 6

    (Hint: See the Chapter 22 supplemental diagrams.)

2. In a nuclear power plant, fission of what isotope produces steam?
    (a) Uranium–235
    (b) Uranium–238
    (c) Thorium–232
    (d) Plutonium–239

    (Hint: See the Chapter 22 supplemental diagrams.)

3. The Lander Oil Field is in the middle of:
    (a) An anticline
    (b) A plunging anticline
    (c) A syncline
    (d) A plunging syncline

    (Hint: See Slide 10.7 in the second edition Slide Set.)

4. Most of the coal swamps for which the Carboniferous is famous appeared during the:

  (a) Early Paleozoic era
  (b) Mississippian period
  (c) Pennsylvanian period
  (d) Cretaceous period

(Hint: See the Geological Time Scale tool.)

FINAL REVIEW

After reading this chapter, you should:

- Understand the profound importance of energy to existence
- Know how and where oil and gas form
- Know something about the environmental and geopolitical concerns relating to oil and gas
- Know how and where coal is mined, as well as some of the environmental concerns relating to coal
- Know about oil shale and tar sands as energy sources
- Know something about nuclear energy and the problems posed in the disposal of radioactive waste
- Know something about the potential of solar and geothermal energies
- Understand the importance of conservation and energy policy

ANSWER KEY

TERMS AND CONCEPTS

1. Nuclear, 2. Oil shale, 3. Wood, 4. Coal, 5. Natural gas, 6. Anticlinal trap, 7. Alberta, Canada, 8. The Sun, 9. Geothermal energy

PRACTICE MULTIPLE-CHOICE QUESTIONS

1. b, 2. c, 3. a, 4. d, 5. a, 6. c, 7. b, 8. d, 9. a, 10. b, 11. a, 12. d, 13. c, 14. a

23

Mineral Resources from the Earth

L ike fossil fuels, minerals are vital to the functioning of a modern nation. Just about everything we use comes from the ground. This chapter surveys a broad range of Earth's mineral resources: their geological settings and their economic and social context, including issues of distribution, depletion, environmental and health costs, recycling, substitution, and price.

◆　　　The chemical elements of Earth's crust are widely distributed in many kinds of minerals, and those minerals are found in a great variety of rocks. In most places, any given element will be found homogenized with the other elements in amounts close to its average concentration in the crust. Elements that occur in higher concentrations have undergone some geologic process that has segregated much larger quantities of the element than normal. High concentrations of elements are found in a limited number of specific geological settings. These are the settings of economic interest.

Rich deposits of minerals from which valuable metals can be recovered profitably are called ores; the minerals containing these metals are ore minerals. Ore minerals include sulfides, oxides, and silicates. Also, some metals, such as gold, are found in their native state—that is, uncombined with other elements.

The concentration factor of an element in an ore body is the ratio of the element's abundance in the deposit to its average abundance in the crust. The economical concentration factor varies from element to element, depending on its average abundance.

Current estimates indicate that minerals will be available and affordable for the next century or so. Rates of mineral consumption in the United States and other industrialized nations have slowed, and the demand for metals, in particular, has been reduced by recycling the metal content of discarded goods and by substitution of low-cost ceramics, composites, and plastics. Mineral resources are ultimately finite, however, and it is wise to think about the long-term future.

◆   Mineral deposits are created by various geologic processes. In general, a mineral deposit forms when a source of the minerals exists in a place where it is accessible to a natural transport mechanism; a natural transport mechanism is available to move the minerals away from the source; and a site exists with a mechanism for the transport agent to deposit the minerals.

Many of the richest known ore deposits crystallized out of hydrothermal solutions. These hot waters can emanate directly from the magma of an igneous intrusion and carry away in solution ore constituents derived from the soluble components of the magma. Hydrothermal solutions can also form when circulating groundwater contacts heated rock or a hot intrusion, reacts with it, and carries off ore constituents released by the reaction.

Ore constituents are often deposited in fractured rocks. The hot fluids flow easily through the fractures and joints, cooling rapidly in the process and precipitating the ore constituents. The tabular or sheetlike deposits of precipitated minerals in the fractures and joints are called vein deposits, or simply veins. Some ores are found in veins; others are found in the rock adjacent to the veins (country rock) that was altered when the hot solutions heated and infiltrated them.

Mineral deposits that are scattered through volumes of rock much larger than veins are called disseminated deposits. In igneous and sedimentary rocks, minerals are disseminated along abundant cracks and fractures.

◆   The most important igneous ore deposits are found as segregations of ore minerals near the bottom of intrusions. The deposits are formed when minerals crystallize from molten magma, settle, and accumulate on the floor of a magma chamber. Most of the chromium and platinum ores in the world are found as layered accumulations of minerals that formed in this way. Pegmatites are extremely coarse-grained intrusive rocks of granitic composition that are usually found as veins, dikes, or lenses in granitic batholiths. They form by fractional crystallization of a granitic magma. Pegmatites may contain rare mineral deposits rich in such elements as boron, lithium, fluorine, niobium, and uranium and in gem minerals such as tourmaline. Diamonds occur chiefly in ultramafic igneous rocks called kimberlites. These rocks were forcefully intruded to the surface from deep in the crust and upper mantle in the form of long, narrow pipes.

◆   Sedimentary mineral deposits include some of the world's most valuable mineral sources, including limestones, pure quartz sands, coarse sand and gravel, clays, evaporites, and phosphate rocks. Sedimentary mineral deposits are also important sources of copper, iron, and other metals. These deposits were chemically precipitated in sedimentary environments to which large quantities of metals were transported in solution. Many rich deposits of gold, diamonds, and other heavy minerals such as magnetite and chromite are found in placers, ore deposits that have been concentrated by mechanical sorting by river currents.

♦   With the advent of plate-tectonics theory, the various types of igneous activity could be explained in terms of the interactions at plate boundaries. *(See the accompanying article on ore deposits and plate tectonics.)*

ORE DEPOSITS AND PLATE TECTONICS

In 1979, geologists exploring the seafloor at the East Pacific Rise found mineral-rich hot springs venting on the seafloor. These hot springs have their origin in seawater that circulates through fractures near the rift where the plates separate. The seawater is heated when it comes in contact with magma or hot rocks deep in the crust. The water leaches minerals from the hot rocks and rises to the seafloor, where the minerals precipitate. Enormous quantities of sulfide ores rich in zinc, copper, iron, and other metals are being deposited along mid-ocean spreading centers.

Soon geologists began to look on land for the remains of ancient seafloor, which might also hold valuable resources. Some deposits were found in plate-collision zones where fragments of ancient oceanic lithosphere (ophiolites) are occasionally emplaced on land. The rich copper, lead, and zinc sulfide deposits in the ophiolites of Cyprus, the Philippines, the Apennines in Italy, and elsewhere probably owe their origin to hydrothermal circulation along ancient mid-ocean rifts.

Many other types of sulfide ore deposits, of hydrothermal or igneous origin, are found at modern and ancient plate-collision boundaries.

Figure 23.15 summarizes some of the associations between plate tectonics and mineral deposits.

Deposits found in magmatic arcs are thought to result from the igneous activity that typically occurs in collision zones. Manganese nodules—potato-shaped aggregates of manganese, iron, copper, nickel, cobalt, and other metal oxides—are found on the seafloor away from plate boundaries. They are formed by the precipitation of these metal oxides from seawater, usually on a small nucleus such as a shark's tooth or a chip of rock. They are potentially valuable because of the gradual depletion of high-grade deposits of manganese on land and because they are rich in other metals.

This brief summary of the geology of mineral deposits barely touches on the great diversity of geological settings in which various minerals of value are found. Although there is probably an abundance of ore bodies on the deep seafloor, most known ore bodies are found on the continental crust. They either originated on the continent or occur as remnants of mineralized pieces of ocean crust thrust onto the continent in plate collisions. Table 23.2 shows the geologic occurrence and uses of some of the main mineral deposits.

◆   Although the growth of mineral demand has now declined in the industrialized nations, unequal sharing of world mineral resources still remains and has far-reaching economic and political repercussions. International relations are deeply affected by struggles over the control of resources.

As the Earth's human population grows and people in less developed nations seek higher standards of living, the demand for mineral resources will be strong. To meet this demand, we will probably need a combination of conservation, recycling, and more efficient use of materials; increased discovery and the "clean" exploitation of mineral and energy resources; and development of low cost, environmentally benign substitutes for minerals, such as fiber-optic cables made of glass instead of copper wire. Ultimately, however, science will be of no avail unless population growth is stabilized. The world clearly faces major readjustments in the decades ahead, and it is certainly not too soon to start working out equitable, humane, and lasting approaches to building a sustainable world.

CHAPTER 23 OUTLINE

PRACTICE MULTIPLE-CHOICE QUESTIONS

Circle the option that best answers the question.

1. The concentration factor for a mineral is defined as:
    (a) The ratio of an element's abundance in a deposit to its average abundance in the crust
    (b) The ratio of an element's abundance in a deposit to its average abundance in the crust and upper mantle
    (c) The ratio of an element's abundance in a deposit to its average abundance in the entire Earth
    (d) Its concentration as measured against other minerals found in the immediate area of the deposit being examined

2. Which of the following minerals has the lowest concentration factor?
    (a) Aluminum
    (b) Nickel
    (c) Lead
    (d) Mercury

3. Which of the following minerals has the highest concentration factor?
    (a) Aluminum
    (b) Nickel
    (c) Lead
    (d) Mercury

4. Strategically critical metals include:
    (a) Lead, silver, and gold
    (b) Zinc, gold, and platinum
    (c) Iron, copper, and nickel
    (d) Manganese, chromium, and titanium

5. Hydrothermal vein deposits would most likely be found in the vicinity of:
    (a) Intrusive plutons
    (b) Shallow-water marine deposits
    (c) Low-grade metamorphism
    (d) Extrusive basalt flows

6. The copper deposits of the southwestern United States are examples of:
    (a) Hydrothermal vein deposits
    (b) Disseminated deposits
    (c) Massive plutons
    (d) Igneous ore deposits

7. Examples of rich igneous ore deposits include the:
   (a) Copper deposits of Chile
   (b) Manganese nodules on the ocean floor
   (c) Sulfide ore deposits of Sudbury, Ontario
   (d) Iron deposits around Lake Superior

8. Placer deposits tend to collect in areas where:
   (a) Winds have blown away the less dense minerals
   (b) Water has removed the less dense minerals
   (c) Heavy minerals settle out of magmatic sources
   (d) Plate tectonics has been very active in moving land masses

9. Iron ore would be least likely to be associated with:
   (a) Plutonic rocks
   (b) Sedimentary layers
   (c) Placer deposits
   (d) Deposits concentrated by groundwater

10. Which of the following are not sedimentary ore deposits?
    (a) Kimberlites
    (b) Evaporites
    (c) Banded iron ores
    (d) Gold and silver placers

11. Geologists first found mineral-rich hydrothermal vents:
    (a) At the Mid-Atlantic Ridge
    (b) At the East Pacific Rise
    (c) At Yellowstone National Park, Wyoming
    (d) On the floor of the Indian Ocean

12. Manganese nodules contain rich deposits of:
    (a) Manganese, iron, gold, and zinc
    (b) Manganese, iron, copper, and zinc
    (c) Manganese, iron, copper, and nickel
    (d) Manganese, tin, copper, and zinc

13. Which of the following statements is true?
    (a) Most ore bodies are found on the continents
    (b) Most ore bodies are found on the seafloor
    (c) Ore bodies are roughly equally distributed between the seafloor and the continents
    (d) Most ore bodies are found at transform fault boundaries

CD RESOURCES

SUPPLEMENTAL DIAGRAMS

23.18   Crustal abundance of economically important elements
23.19   Percentage and value of nonfuel minerals produced and consumed
23.20   Cycle of mining, preparation, use, and discard of useful Earth materials
23.21   The gold pan, an old prospecting tool
23.22   Role of plate boundaries in the accumulation of mineral deposits
23.23   Polymetallic sulfide deposits at Pacific Ocean spreading centers

SUPPLEMENTAL PHOTOS

23.24   Percentage and dollar value of nonfuel minerals produced and consumed

FIRST EDITION SLIDE SET

Slide 23.1   Gold and platinum nuggets
Slide 23.2   Fluid inclusions in a gold-bearing quartz vein
Slide 23.3   Sulfide-bearing quartz vein, Sierrita Mine, southeastern Arizona
Slide 23.4   Vein of copper sulfide ores, Santa Eulalia, Mexico

ILLUSTRATED GLOSSARY

EXERCISES

Plate Boundaries and Mineral Deposits

CD LAB

1. Which economically important mineral is most abundant in the Earth's crust?
   - (a) Iron
   - (b) Aluminum
   - (c) Magnesium
   - (d) Potassium

   (Hint: See the Chapter 23 supplemental diagrams.)

2. Massive sulfide mineral deposits are found in ophiolites:
   - (a) On the west coast of North America
   - (b) On the east coast of North America
   - (c) On the west coast of South America
   - (d) On the east coast of South America

   (Hint: See the Chapter 23 supplemental diagrams.)

3. Placer gold typically contains:
   - (a) 5 to 10 percent gold
   - (b) 10 to 20 percent gold
   - (c) 10 to 25 percent gold
   - (d) 15 to 20 percent gold

(Hint: See the first edition Slide Set.)

4. The percentage of gold in 14-karat gold is about:
   - (a) 20 percent
   - (b) 40 percent
   - (c) 60 percent
   - (d) 80 percent

(Hint: See the first edition Slide Set.)

FINAL REVIEW

After reading this chapter, you should:

- Understand the economic importance of mineral deposits
- Know how hydrothermal deposits form
- Know the differences between igneous ore deposits and sedimentary ore deposits
- Understand the relationship between ore deposits and plate tectonics
- Know some of the issues confronting society relating to mineral resources

ANSWER KEY

PRACTICE MULTIPLE-CHOICE QUESTIONS

1. a, 2. a, 3. d, 4. d, 5. a, 6. b, 7. c, 8. b, 9. c, 10. a, 11. b, 12. c, 13. a

24

Earth Systems and Cycles

Throughout, the textbook has emphasized the interdependence of Earth's internal and external processes. This chapter looks at how all Earth's major systems—the interior, the lithosphere, the atmosphere, the hydrosphere, and the biosphere—are connected. Past changes on Earth can be linked to geological perturbances of its systems and cycles, and there is reason to fear that human-induced perturbances could cause global catastrophe in the future.

◆ Earth's internal systems created its external systems. Outgassing during differentiation and volcanism provided Earth with an atmosphere and a hydrosphere, although oxygen did not enter the atmosphere until photosynthetic organisms evolved.

Much of the Sun's radiant energy is absorbed by Earth's surface and then radiated back to the atmosphere as invisible infrared (IR) heat rays. Because carbon dioxide and water molecules strongly absorb the infrared instead of allowing it to escape into space, the atmosphere is heated and radiates heat back to the surface. This process is called the greenhouse effect because it is like the warming of a greenhouse, whose glass lets in visible light but lets little heat escape. The more carbon dioxide, the warmer the atmosphere; the less carbon dioxide, the colder. Without the greenhouse effect, Earth's surface temperature would be well below freezing and the oceans would be a solid mass of ice.

◆ The earliest evidence of life is microscopic fossils of bacteria, probably marine, at least 3.5 billion years old. These bacteria are though to have evolved from large organic molecules assembled from gases such as methane and ammonia. The organic molecules aggregated and formed a general system for growth and metabolism. This kind of system is thought of as protolife, because it was not self-reproducing. The next step would have been the development of the first truly self-replicating molecule, RNA. These early steps in the origin of life probably did not significantly affect the early atmosphere, which remained dominated by nitrogen as well as carbon dioxide. But the next evolutionary development, the arrival of photosynthesis, did profoundly affect the atmosphere, the hydrosphere, and the lithosphere.

◆   Photosynthesis is the process by which green organisms use chlorophyll and the energy from sunlight to make carbohydrates out of carbon dioxide and water. Because animals cannot synthesize carbohydrates for themselves, they depend on food from plants or other animals for the energy they need to live. To retrieve that stored energy, the organism takes in oxygen and, in its cells, combines oxygen with carbohydrate. The chemical reaction that releases the energy stored by photosynthesis is called respiration. Figure 24.3 shows how photosynthesis and respiration are reciprocal processes.

Geological evidence indicates that by 1.5 billion years ago, there was a significant amount of oxygen in the atmosphere. The explosion of life forms in the late Precambrian and Cambrian was probably stimulated at least in part by a rise in oxygen levels to levels comparable to today's. Table 24.1 summarizes significant developments in the evolution of our planet and the life on it.

◆   Earth changes because chemical elements such as carbon and oxygen circulate through it. Photosynthesis and respiration are part of the cycle that circulates carbon through the atmosphere, biosphere, and hydrosphere. Geologists have found that geochemical cycles such as the carbon cycle maintain the Earth in more or less a steady state. Earth's atmosphere, hydrosphere, and other systems can be viewed as reservoirs for holding Earth's chemicals. These reservoirs are linked by transport processes between them. Geochemical cycles trace the flow, or flux, of Earth's chemicals from one reservoir to another.

◆   The calcium cycle is one important geochemical cycle. Among many other elements, the ocean contains a large amount of calcium, which enters steadily in large quantities through rivers. If the ocean kept receiving this much calcium with no calcium outflow, it would quickly become supersaturated with calcite and gypsum. The flux that removes the majority of excess calcium from the ocean is the sedimentation of calcium carbonate. The inflow of elements into the ocean is approximately equal to the outflow, so the oceans are close to a steady state.

◆   Figure 24.9 shows the fluxes and reservoirs that govern the carbon cycle. The carbon cycle keeps carbon in Earth's reservoirs balanced over the long term, but over the short term the burning of fossil fuels is adding carbon dioxide to the atmosphere at a fast rate, threatening to affect global climate and the biosphere.

◆   Figure 24.10 shows a simplified geochemical model of the Earth. In this steady-state model, all fluxes are balanced. This balance sustains life as we know it on Earth. If the current balance were to be greatly perturbed, many of Earth's life forms would face devastation and perhaps extinction.

CLIMATE CHANGE AND PLATE TECTONICS

Geologists are investigating several hypotheses that link plate tectonics to climate change. One new idea explains the climatic cooling and growth of ice sheets during the Cenozoic era by proposing a sequence of events that began with the collision of the Indian and Eurasian plates and the uplift of the Himalayas and the Tibetan Plateau. The Tibetan Plateau grew to a height of 5 km and an area half the size of the United States to become a dominant feature of Earth's topography. The scientists behind this new idea propose that the growth of the plateau would have had a profound effect on the monsoons of the region. Monsoons are major wind systems that reverse direction seasonally. They typically blow from cold to warm regions—from sea toward land in the summer and from land toward sea in the winter. In southern Asia, the spring and summer heating of the Tibetan Plateau causes the air above to rise, creating a low-pressure region that draws in moisture-laden air from the adjacent oceans. The resulting winds rising over the southern slopes of the Himalayas bring heavy rains that swell the rivers.

The combination of heavy monsoon rains and high relief accelerates physical erosion and chemical weathering of the newly exposed silicate rocks in the mountains. The carbonic acid in rainwater (formed by the solution of atmospheric carbon dioxide in water) accelerates the chemical breakdown of silicate rocks. The weathering products, which include minerals that lock up the carbon dioxide, are flushed away by streams, which constantly exposes fresh minerals to chemical attack. This reaction results in a reduction of atmospheric carbon dioxide, a reduced greenhouse effect, and global cooling. Computer models of the growth of the Tibetan Plateau and climatic change seem to support this new hypothesis linking plate tectonics and the external environment.

Other mechanisms that link plate tectonics and global climate are also being researched. These include sea-level changes, the drift of continents over the poles, and tectonic movements that block ocean currents or open gateways for currents to flow through. For example, the emergence of the Isthmus of Panama closed a passage connecting the Atlantic and Pacific oceans. If future geologic activity were to close the narrow channel between the Bahamas and Florida through which the Gulf Stream flows, temperatures in western Europe would drop drastically.

♦   Since life began, the course of biological evolution has both caused and been affected by changes in the atmosphere and hydrosphere—the systems primarily involved in weather and climate. The geologic record suggests that long-term swings in global climate have occurred several times in Earth's evolution. To explain these climate changes, geologists are considering several hypotheses that link plate tectonics with the external environment. *(See the accompanying article on climate change and plate tectonics.)*

Paleontologists have found dozens of mass extinctions in the fossil record. Global climate change is a favored explanation, and several causes of such change have been proposed, including superplumes (huge eruptions of flood basalts), impacts from bolides (asteroids and comets), and intense radiation from a nearby exploding star. The greatest mass extinction occurred 250 million years ago at the boundary between the Permian and Triassic periods, referred to as the P-T boundary, when some 90 percent of marine species and 70 percent of land vertebrates were wiped out. The geologic record shows a sea-level drop, an ice-sheet expansion, and an interval of acid rain at this time. The second-greatest mass extinction occurred about 65 million years ago, at the boundary between the Cretaceous and Tertiary periods, referred to as the K-T boundary. Some scientists are convinced that volcanism was the cause and cite as evidence the major superplume eruption of flood basalts that occurred in India at that time. The most widely accepted hypothesis now, however, is that a giant bolide slammed into Earth, releasing 4 billion times more energy than the Hiroshima atomic bomb.

The bolide impact hypothesis has inspired much debate and research. Evidence lies in a large crater on the Yucatan Peninsula of Mexico, together with a layer of breccia near the top of the Cretaceous strata containing shock-metamorphosed crystals of quartz and feldspar and rock fragments melted into glass. Shock metamorphism occurs when minerals are subjected to the high pressures and temperatures of shock waves generated by impacts. But the most convincing evidence for the bolide impact hypothesis is a high concentration of the element iridium in the K-T boundary layer. Iridium is the chemical fingerprint of a bolide impact because the element is abundant in bolides but rare in Earth's crust.

♦   The term *global change* entered the world's vocabulary as evidence mounted that emissions from human activities could alter the chemistry of the atmosphere with disastrous worldwide consequences. These include climate change due to an enhanced greenhouse effect, increased exposure to ultraviolet rays because of stratospheric ozone depletion, mass die-offs due to acid precipitation, and an overburdening of many Earth systems due to overpopulation

The burning of fossil fuels releases carbon dioxide and other gases into the atmosphere. At the present rate at which we burn fossil fuels and destroy forests, the amount of carbon dioxide in the atmosphere may reach double the preindustrial level in the second half of the next century. The combined effect of emissions of carbon dioxide and other gases could be a global warming of 1 to 3.5°C toward the end of the next century. Global warming could cause changes in wind and rainfall patterns and soil moisture that could convert some of Earth's most productive agricultural regions into semiarid wastelands. Because human-induced warming would be more rapid than any that occurs naturally, many plant and animal species would have difficulty adjusting or migrating. Those that could not cope with rapid warming would become extinct. Along with changes in wind and rainfall, the oceans could warm and expand, raising the sea level as much as 60 cm, a serious problem in low-lying countries such as Bangladesh. If the continental ice sheets start to melt, the sea level could rise even higher. *(See Expansion Module Chapter 34 on the CD, "The Rising Seas.")*

Ozone in the stratosphere forms a protective layer around the Earth that absorbs certain portions of cell-damaging ultraviolet radiation. Skin cancer, cataracts, impaired immune systems, and reduced crop yields are attributable to excessive ultraviolet exposure. In 1996, the Nobel Prize in chemistry was awarded to Sherwood Rowland and Mario Molina for the hypothesis they advanced 22 years earlier that the protective ozone layer can be depleted by a reaction involving human-made compounds called chlorofluorocarbons (CFCs). CFCs, which are used as refrigerants, propellants, and cleaning solvents, are stable and harmless except when they migrate to the stratosphere. There, the intense sunlight breaks down CFCs and releases their chlorine, which reacts with the ozone molecules in the stratosphere and thins the protective ozone layer.

As we saw in Chapter 22, the burning of high-sulfur coal releases gases such as sulfur dioxide. Sulfur gases in the atmosphere react with oxygen and rainwater to form sulfuric acid. Some nitric acid is formed in the same way from nitrogen oxide gases emitted from smokestacks and automobile exhausts. Small amounts of sulfuric and nitric acids turn harmless rainwater into acid rain, which causes widespread damage to water and air, delicate organisms, and solid rock.

The world population of 5.7 billion in 1995 is expected to double by 2050. The developed countries, with only 23 percent of the world's population, produce 85 percent of the world's economic output and withdraw the majority of the minerals and fossil fuels from Earth's finite resources. The use of these resources is largely responsible for today's level of pollution. Uncontrolled use of resources is a serious and growing encroachment by humankind on the interacting systems of Earth's atmosphere, hydrosphere, and land surface.

◆   The world faces several dilemmas. How do we reduce greenhouse gases and other pollutants without impeding economic growth? How can we meet the legitimate aspirations of the developing countries for economic growth without adding to the planet's load of pollutants? The population explosion in the developing countries compounds the problem of how to sustain both economic growth and a healthy environment. Unlimited population growth would lead to profound changes in the global environment. Human health would be threatened. The world would suffer an irreversible loss of biodiversity (the world array of wildlife and plants) with large-scale devastation of plant and animal habitats. There would be political unrest and mass migrations of people.

Is it possible to have sustainable development in both the developed and developing nations—that is, economic growth and improving standards of living that can last indefinitely and are environmentally benign? The stabilization of population growth and advances in science and technology might allow us to avoid a calamity as serious as any humankind has ever faced.

CHAPTER 24 OUTLINE

A THEORY OF EARTH'S EVOLUTION

Fill in the blanks to complete the table below.

YEARS AGO	EVENT
15 billion	The Big Bang brings the universe into being
1. _____	Our solar system takes shape
4.5–4 billion	2. _____
	Early atmosphere forms
3. _____	Organic molecules synthesize and stabilize in the sea
	Life begins with the evolution of self-replicating molecules
4. _____	Bacteria and photosynthetic unicellular organisms evolve
	5. _____
1.5–0.5 billion	Multicellular marine organisms evolve
6. _____	Complex marine organisms evolve
0.45 billion–present	Land plants and animals evolve

PRACTICE MULTIPLE-CHOICE QUESTIONS

Circle the option that best answers the question.

1. Earth's _____ systems created its _____ systems.
    - (a) External, internal
    - (b) Internal, external
    - (c) Atmospheric, internal
    - (d) Hydrospheric, external

2. Carbon dioxide and water:
    - (a) Strongly absorb infrared radiation
    - (b) Weakly absorb infrared radiation
    - (c) Strongly reflect infrared radiation
    - (d) Weakly reflect infrared radiation

3. The earliest evidence of life is:
     (a) 100 million years old
     (b) 1 billion years old
     (c) 2.5 billion years old
     (d) 3.5 billion years old

4. Photosynthesis and respiration are:
     (a) Equilibrium processes
     (b) Self-regulating processes
     (c) Parallel processes
     (d) Reciprocal processes

5. Monsoons typically blow:
     (a) From the sea toward land in winter
     (b) From the sea toward land in summer
     (c) From land toward the sea in summer
     (d) In the same direction all year

6. Which of the following is *not* considered a possible cause of global climate change that caused mass extinctions?
     (a) Bolide impact
     (b) Plate tectonics
     (c) Intense radiation from a nearby star
     (d) Superplumes

7. The greatest mass extinction of all time occurred:
     (a) At the Permian–Triassic boundary
     (b) At the Cretaceous–Tertiary boundary
     (c) At the Phanerozoic–Paleozoic boundary
     (d) At the Pliocene–Pleistocene boundary

8. The 1996 Nobel Prize was awarded for:
     (a) The theory of plate tectonics
     (b) The theory of the "greenhouse effect"
     (c) The theory of ozone depletion by CFCs
     (d) The theory of bolide impacts

9. The 1995 world population of 5.7 billion is expected to double by the year:
     (a) 2000
     (b) 2010
     (c) 2025
     (d) 2050

CD RESOURCES

SUPPLEMENTAL DIAGRAMS

24.21 Feedback system that keeps ocean approximately saturated with calcium carbonate

24.22 Production of acid rain by coal combustion

SECOND EDITION SLIDE SET

24.1 Fossil leaf of *Glossopteris* from the upper Permian Dunedoo Formation in southwestern Australia

24.2 Aerial view of Barringer Meteor Crater, Arizona

24.3 Dinosaur tracks in northern Arizona

Also see Slides 7.1, 7.2, 15.2, 15.3, 15.4, 15.5, 15.6, and 15.7.

EXPANSION MODULES

Chapter 34 "The Rising Seas" (*Scientific American* article)

ILLUSTRATED GLOSSARY

EXERCISES

The Carbon Cycle

CD LAB

1. How should we expect the Antarctic ice cap to change as a result of global warming?
    (a) It will probably shrink, resulting in a rise in global sea level
    (b) It will probably expand, resulting in a fall in global sea level
    (c) It will probably remain the same
    (d) Nobody really knows

   (Hint: See Expansion Module Chapter 34.)

2. Most researchers agree that sea level is currently:
    (a) Rising
    (b) Falling
    (c) Not changing

(Hint: See Expansion Module Chapter 34.)

Web Project: One of the interesting links on the Chapter 24 page is the U.S. Census Bureau, which provides "population clocks" that show up-to-the-minute population projections for the United States and the world. According to the U.S. clock, there is a net gain in the United States of one person every 12 seconds—a statistic that should give us pause Another link is Simulation of Global Warming, which contains animations of 70-year predictions for global surface warming and sea-ice coverage.

FINAL REVIEW

After reading this chapter, you should:

- Be able to describe how the atmosphere and the hydrosphere originated
- Know how the process of photosynthesis altered the atmosphere
- Be able to describe the calcium cycle and the carbon cycle
- Be able to propose and describe two causes for mass extinctions
- Understand how greenhouse gases affect global warming
- Understand the consequences of unlimited population growth

ANSWER KEY

A THEORY OF EARTH'S EVOLUTION

1. 4.6 billion,   2. Earth accretes, melts, and differentiates,   3. 4–3.5 billion, 4. 3.5–1.5 billion,   5. Oxygen from photosynthesis changes the atmosphere, 6. 0.5 billion–present

PRACTICE MULTIPLE-CHOICE QUESTIONS

1. b, 2. a, 3. d, 4. d, 5. b, 6. b, 7. a, 8. c, 9. d